THE PHILOSOPHY OF HUMAN MOVEMENT

First Edition

By Dina Mijacevic

Washington State University

Bassim Hamadeh, CEO and Publisher
Michael Simpson, Vice President of Acquisitions
Jamie Giganti, Managing Editor
Jess Busch, Senior Graphic Designer
Kristina Stolte, Acquisitions Editor
Michelle Piehl, Project Editor
Alexa Lucido, Licensing Coordinator
Mandy Licata, Interior Designer

First published in the United States of America in 2016 by Cognella, Inc.

Printed in the United States of America

ISBN: 978-1-63487-379-6 (pbk) / 978-1-63487-380-2(br)

www.cognella.com 800-200-3908

DEDICATION

I dedicate this text to my mentor, Dr. Sharon Kay Stoll. She is one of the most influential sport people in the world who taught me the true meaning of movement and physical activity. Every time my mind begins to wander and gets disconnected from my body, I think about her philosophical insights. She had shared her movement experiences and managed to make sure I succeeded on my own, providing encouragement in every physical activity and movement story. I cannot say enough to thank you for everything you have done. You do make a difference in every student's college experience, and some day, I wish I could be like you.

CONTENTS

PREFACE

Movement professionals often work in very hectic environments with professional, recreational, high school, collegiate, athlete, and master athletes, treating injuries, preventing injuries, and working on their top performance. It is well known that many individuals value the benefits of physical activity, but on the other hand, many of us chose not to be active.

Movement sciences majors are usually undervalued and considered to be the "ghetto" of the majors. Physical education programs are usually the first ones to be cut from university majors and public schools. It is clear that intellectual subjects, like math, chemistry, and biology, are viewed as more important than movement programs. Hence, mind is valued over body, body is subservient to mind, and thinking is more important than doing.

I hope this book helps you realize that movement is important and significant to being human. The essential value of movement and physical activity is imperative for our personal growth and success. Physical activity was once required for human existence, but now, it just seems to be elective.

As movement professionals, we need to understand why some paths to physical activity work and why some do not. Simply said, often times we prescribe movement based on what is good for the body, not necessary what is good for the individual. For some people, movement and physical activity can be challenging and stimulating. Running on a trail can be torturous; friendly competition can turn best friends into enemies. Young athletes pushed too hard can completely lose interest in movement for the rest of their lives.

When we find that intimate relationship with movement, it can be aesthetic, peaceful, liberating, and engaging. We need to examine movement experiences that lead to these positive relationships with physical activity.

PURPOSE OF THE TEXT

This text is for kinesiology and movement sciences students interested in a different approach to physical activity. It is important to mention that I do not have a background in movement philosophy or philosophy in particular. Advanced kinesiology students, experienced movement professionals, and well-educated sport and exercise professionals should use this text.

In addition, this book can be a good resource for graduate students interested in getting familiar with philosophy's role in movement programs driven more toward scientific and empirical practice. Graduate and undergraduate students are encourage to use this text in conjunction with R. Scott Kretchmar's book *Practical Philosophy of Sport and Physical Activity.* This text may help you realize the true meaning of physical activity and movement in general. It may also help you discover who you are as a moving individual or who you are striving to be. In fact, it may help you understand that physical activity and movement can bring happiness and satisfaction in one's life.

However, I am also aware that people have different upbringings and, therefore, different opinions and that there are many answers to one subject. This text will by no means answer all of your questions or provide a solution to all of your problems. R. Scott Kretchmar (2005) summed it well by saying that much like science, philosophy is never finished. We work hard to solve problems, and when we think we have discovered something new and remarkable, another opportunity presents itself. This text is about the journey, not the destination.

OBLIGATIONS AND EXPECTATIONS

In this text, I borrowed from many philosophic ideas. Most of the theories found in this text come from R. Scott Kretchmar's *Practical Philosophy of Sport and Physical Activity* (2005). I also borrowed some philosophic movement and physical activity arguments from my mentor, Dr. Sharon Kay Stoll from the University of Idaho Center for Ethics. I am not a sport ethicist, but I was fortunate enough to study sport ethics with one of the leading ethicist in the United States. The ethics chapter was developed from everything I have learned from Dr. Sharon Kay Stoll. In addition, Maurice Merleau-Ponty and Edmund Husserl supported some of my arguments, because their philosophy is the closest to the mind/body connection. Most of my positions were created through my movement experiences and conversations I have shared with my Human Movement Philosophy (KINES 314) students at Washington State University.

I must say that this text favors certain opinions like mind/body holism over mind/body dualism. My intention is not to have my readers choose one over the other, but to have a background in both. Also, I am a big advocate of meaningful movement experiences. In fact, there are many things in our lives that have meanings, but not all of them are meaningful. I guess I would have to say that I am a little biased. One of my professors in graduate school once told me that it is biased not to have a bias. Often times, I do think that we can solve problems and find solutions through movement and physical activity.

By the end of this text, I can only hope that you will be able to find some guidance to help guide you through your future careers in movement and physical activity. Maybe you will be able to find some personal answers to why you value movement and physical activity.

I do think that our experiences are more important than any scientific discoveries. It does not matter how we develop those experiences, but our willingness to share them through their movement stories. Dr. Sharon Kay Stoll once said that it is often through movement, whether it's a recreational game or some sporting event, where we can see through people's souls.

ORGANIZATION AND STRUCTURES

At the beginning of each chapter, you will see a brief introduction of each topic. In addition, each chapter contains some intriguing/mind-provoking questions. Thinking activities will interrupt most of the chapters from time to time. These activities provide an opportunity to form your own conclusions on certain topics. Through these thinking activities, you will practice developing your own philosophic skills. As R. Scott Kretchmar (2005) would say: "Philosophy asks you to slow down, think for yourself, and intelligently evaluate your personal and professional options" (p. xiv).

This book is organized into 10 chapters. The first chapter describes what philosophy is and how it can be used in movement professions. Chapter 2 and 3 explain mind/body dualism and holism and their application to physical activity and movement. Both mind/body dualism and holism have their strengths and weaknesses. I provide examples and arguments for both. We explore why holism may be more effective for movement professionals.

Chapters 4 and 5 go deeper into the passion behind movement and physical activity, explaining what that meaning is and why we need it to understand movement. Chapter 6 introduces professional ethics, moral, and non-moral values, what is a professional, and why we need to practice good professional ethics.

Chapter 7 examines the importance of extrinsic and intrinsic values. We will explore the importance of fitness/health, fun/pleasure, and movement skill/knowledge. Chapter 8 distinguishes between games and play and their significance. Chapter 9 explores competition and winning, the importance of winning, and the importance of playing fair.

Chapter 10 goes into explaining self-knowledge and discovery and what constitutes and counts as knowledge. Chapter 11 asks us what are we promoting through movement professions and what is our movement and physical activity philosophy. I truly hope that you will take this opportunity to reflect on your experiences and let them guide you through your own personal movement/physical activity philosophy.

CHAPTER ONE

Philosophic Skills in Movement Activity

In today's performance and science–oriented culture many kinesiology students are faced with struggles to defend their profession. Physical education programs at universities have changed to have more scientific-based names like movement sciences. The incorporation of science is, in itself a benefit for the discipline; however, it has also led to decreasing subjective experiences in the joy in movement. Many educators believe that liberal arts and individual experiences in kinesiology provide for more enrichment than essential knowledge, particularly in preparing for a professional career in health and movement. In fact, in 1963 Harvard president James Bryant Conant claimed that graduate programs in physical education are an example of the system at its worst. The physical education programs had become "just a gym class" with the attendance of "dumb jocks" (Anderson, 2002). Simply put, the movement programs without sciences are predominately the university ghettos. The above claim by James Bryant Conant had led the American Academy of Kinesiology and Physical Education to approve the name kinesiology for the previously called physical education program. It seems appropriate to call any movement programs kinesiology because in Greek it means "the story or the account of movement" (Slowikowski and Newell, 1990).

Kinesiology and movement sciences programs now emphasize mostly scientific work to obtain respect from other programs. Science programs are more academic and intellectual, and physical education programs are not to be classified as an academic discipline. Due to this argument, a science based program in movement would produce knowledge that could be applied to better athletic performance, bigger and stronger physique, faster recovery time, and fewer injuries. There is less time to introduce anything but science. Science reigns supreme over philosophy, measures real things, and produces facts. Therefore, the place for kinesiology programs would be justified. But, kinesiology still tends to be apologetic toward other academic programs due to its movement-based courses. A stronger emphasis on and a better inclusion of the human side of movement and the experience of movement may be needed for a wider respect. The point of this book is not to disrespect science, but to include valuable arguments, lessons, and practical benefits of human movement. Douglas Anderson's article on the Humanity of Movement: "It's not just a Gym Class" (2002) argues that honoring science to the exclusion of the humanities of movement can have negative consequences, especially if it is inattentive to the creative and disciplinary features of movement experiences that yield personal and social meanings (p. 89).

Movement professionals often get asked questions or hear opinions from clients that are, at times, surprising and, at times, totally vexing. These statements usually are something like this, "I really want to be at my fastest, I think I need to lose 7 pounds by the end of this month," or, "I read that so-and-so can squat 450 pounds, so I'm really focused on leg strength right now," also, "I'm going to cut fat and carbohydrates out of my diet. Ya' know, get leaner" (Mijacevic, 2013). A small sample but indicative of the range of statements one can get to by regarding how to treat their body as an object. The significance and passion for our chosen sport, exercise, physical activity, and movement are consistently undervalued in the majority of studies in physical fitness, exercise physiology, sport psychology, health, and performance. If we cannot quantify, track, develop spreadsheets, and manipulate data in our research studies, we cannot get results. As an outcome, we now know more about lactic acid concentrations than we do about joy for sport, physical activity, exercise, and movement. We know more about body mass index than we do about any significance of movement and sports, more about exercise performance than we do about eagerness and enthusiasm. Simply, we just shred human movement and physical activity down to clean components of anatomy, exercise physiology, and biomechanics and then complain how no one wants to participate (Kretchmar, 1994; Kretchmar, 2005; Forencich, 2006). Just because we cannot measure or quantify something does not mean such information is invaluable. Often, with numbers there are really only two directions: go-higher or lower, slower or faster. Human behavior does not go in only two directions, it goes in many directions. Even when elite and extraordinary athletes and performers, such as LeBron James, Novak Djokovic, and Brittney Griner attract our thirst for their athletic/movement abilities, we study them in terms of data analysis and theories that are already available. We reduce their performance abilities into the tiniest components of laws and statistical descriptions and argue that there is nothing important about their individuality and uniqueness. To know the mechanics of hitting a baseball precisely, but be ignorant of Jackie Robinson's social conflicts, is to be an incomplete kinesiologist.

If scientific measurements can solve our problems, there will be no inactivity and obesity in our society, but also no room for movement philosophy. There will be no focus on movement creativity, meaning, expression, and body language. Thus, an individual interested in movement would be underappreciated in the academic field. If the university worries about weak or uneducated minds, argues Anderson (2002), it should also worry about weak and uneducated bodies (p. 92). It is not enough to read about movement, but we must also participate in movement activities. Human movement contains skills and techniques and allows us to identify ourselves to no longer see ourselves as slaves to a particular activity, but that particular activity becomes we are, and we learn to express our uniqueness through it. Gym classes should be the sites for experiencing movement philosophy and our human excellence.

One important aspect of movement philosophy is to define movement or *kinesis*. Aristotle saw kinesis as a matter of existence significant to human identity as a whole. To be an unique individual is more than a sum of chemicals, but a distinctive whole engaged in the motions of life. The movement is a phenomenon and is the product of subjects, rather than objects. "One thing we know, ideas don't move muscles" (Gerard, 1969, p. 46 as cited in Polanyi, 1969). The importance of movement lies within the mover's experience. In such experiences, "we aim at excellence and virtuosity; we encounter limits and failures; we learn the 'thisness' of movement, not just the theoretical 'how' or the 'what'" (Anderson, 2002, p. 91). This "thisness" of movement is crucial and it cannot be argued, but only felt in the actual lived movement experiences. The magic of alpine skiing down the slope cannot be fully captured by description; whether theoretically or verbal. It must also be experienced-felt! These unique experiences get people hooked on the movement activity and the movement professions. If kinesiology is to be fully redeemed, science must be complemented by the humanities of movement—the features of movement experience that generate, disclose, or develop personal and social meanings and virtues (Anderson, 2002, p. 95). Opposing money-oriented suppositions, science may not always be is not a superior way of knowing. To be an exceptional athlete in any movement activity, one must go through a long course of experiences under the guidance of a master. Therefore, kinesiology should examine physical activity from various perspectives (encompass both pure science and humanities), because games, play, sport, exercise, and dance are central to who we are and what we value.

> Furthermore, it becomes clear that human beings are more than a mere aggregation of chemicals beholden to physical laws. We "make love," "jump for joy," and "sink into depression." We slide "just under the tag," "run to daylight," and "catch our second wind." We are corporeal beings from whom touch is our greatest form of intimacy. We are moving beings that are at home running, jumping, catching, throwing, and kicking. We are embodied beings who dance, play, and compete. We are social, temporal, and communal beings that are born, grow, inherit and belong.
>
> (Twietmeyer, 2012, p. 19–20)

PHILOSOPHIC QUESTIONS

Philosophy is a deliberate and rational attempt to understand the sum and whole of one's experience in both its subjective and objective aspects with a view for effective living. On the contrary, science is a possession of knowledge and is distinguished from ignorance, as a systematic study of knowledge as an object of study, and knowledge attained through study or practice. From these characterizations, we can generate a definition of exercise science as a systematized study and knowledge of facts and theories as applied to sports and those who participate, play, or spectate. Physical education is define as the art and science of human movement as studied and applied to the teaching and coaching professions.

Ask yourself why you have chosen to study human movement, kinesiology, or exercise/ sport sciences.

a. Will your answer hold you in good stead for 10 to 20 years?
b. What are the ramifications of this question to your personal life?
c. Do you play apologetics when you are asked about your major/minor and your future profession?

Ask yourself what intrinsic benefits you get from movement education, administration, coaching, and teaching.

a. Are you committed to your own personal fitness?
b. Are you committed to your own personal play?
c. Are you committed to your own personal learning?
d. Do you get grades to get grades or because you actually want to learn as much as possible?
e. Can you apply what you learned from one class to another or to the real world?

Ask yourself what gifts do you have to give to this field of study.

a. Do you truly like people?
b. Are you committed to supporting fitness for others?
c. Are you committed to supporting play for others?
d. Are you committed to supporting learning for others?

Ask yourself if you have made a commitment to this field of study?

a. In what professional organization are you a member?
b. Do you plan on being a member of any more?
c. Will you be committed to that organization?

PHILOSOPHY IN ACTION (QUESTIONS TO WONDER)

The purpose of philosophy is to help you apply alternate viewpoints, criticize all points of view, understand ramifications, and see the big picture. Examples of philosophic questions are:

- Why do we study the science of sport if those who practice have never studied?
- Why do we study what we study?
- What happens when philosophies collide?
- Should athletics be in the school system?
- Should student teachers be permitted to coach?
- Should physical education be required in public/private schools?
- Should fitness be the goal of physical education programs?

SCHOOLS OF PHILOSOPHY

Idealism

Idealism is a clear distinction between body and mind. Body is important; to be unfit is to be unwise. Athletics are frills and cause one-sidedness, but it is all right if they are a part of curriculum. Idealism is the idea of what sport and movement should be. Athletics are frills … causes one-sidedness, but it is OK if part of curriculum. For example, idealism is the idea of what sport and movement should be.

Realism

Realism is defined as the real world of science. Body is the means though with he sense data. Curricula are only important if they are science-based and tied to laboratory and experimentation. Athletics are only functional under scientific data and outside of curriculum.

Pragmatism

Pragmatism is based on needs, the interests of learners, the variety of activities, and individual choices integrated with the total subject matter of school. Athletics is not favored as specialization, and many would prefer intramurals.

Existentialism

Existentialism embodies the completeness of body and self, curricula by choice, and athletics is only based on individual activities. Individuals have their own choice and responsibility.

It is necessary to mention that these divergent philosophies cause large differences of opinion about professional, personal, and social issues.

PHILOSOPHIC WONDERING IN MOVEMENT

Have you ever wondered how much of you is you? What does it mean to be yourself? Why are you in movement science or kinesiology? Why are you in sport science? Why are you in physical education? Why are you in dance? What will you contribute to exercise science, recreation, sport, and dance? What do you have to contribute to the movement sciences? Who are you in relation to movement? What is your position on required physical education, personal fitness, the joy of movement, the value of play, and the place of movement in lifelong learning?

Philosophy in Greek means love of wisdom (*philia* = love of and *sophy* = wisdom). Greeks argue that there are three type of love: 1) eros or sensual love, 2) agape or unconditional love, and 3) philia or brotherly love. In other words, philosophy is a love of wondering, a love of pondering, love of investigation, and a love of learning. Philosophy is also a deliberate and rational attempt to understand the sum and whole of one's experience in both its objective and subjective aspects with a view for effective living. Through philosophic thinking, clarification and understanding of the nature, purpose, and significance of our very being is facilitated. The philosophic approach is an attempt to place study into a systematized logic pattern so that one can understand how different areas of study function and how they blend together: a) branches of philosophy—the different concepts and areas of study and b) schools of philosophy—how different thinking is applied to the branches of philosophy. For example, movement philosophy is one of the branches of philosophy, which is important and imperative to understanding who you are as a moving individual, and more important, the reason you decided to be a part of the discipline of movement science. The answer is linked to the study of metaphysics—Who you are? What do you believe? Why you chose the discipline may be linked to why you like to play and move. Do you learn better kinesthetically? The answer may be linked to epistemology—How do we know? What do we know? Can we know? Why do you like to watch sport? The answer is linked to axiology and aesthetics. What is your position on cheating? Do you play fair? Do you value fair play? What is sportsmanship? The answer lies in axiology: ethics.

On the contrary, science is explained as a possession of knowledge, as distinguished from ignorance, a systematic study of knowledge as an object of analysis where knowledge is attained through study or practice. Thus, exercise science is a systematized study and knowledge of facts and theories as applied to sport and its participants who play or spectate. Physical education is the art of human movement as studied and applied to the teaching and coaching professions (Metheny, 1972).

METAPHYSICS

Sport philosophy questions may be answered by using the base of the tree of thinking.. Metaphysics derives from the Greek *meta ta physika* (after the things of nature), Metaphysics is an expression used by Hellenistic commentators to refer to Aristotle's united group of texts call the Metaphysics. Classical and medieval philosophers took this title to describe the subjects discussed in the Metaphysics which came after the things of nature, because they were further removed from sense perception and, therefore, more difficult to understand. Since Kant (1993), "metaphysics" has often meant a priori speculation on questions that cannot be answered by scientific observation and experiment. Popularly, "metaphysics" has meant anything abstruse and highly theoretical—a common eighteen century usage illustrated by Hume's (Dicker, 1998) occasional use of "metaphysical" to mean "excessively subtle."

The term has also been popularly associated with the spiritual, the religious, and even the occult. In modern philosophical usage "metaphysics" refers generally to the field of philosophy dealing with questions about the nature of things and their modes of being. Its subject matter included the concepts of existence, objects, property, events; the distinctions between particulars and universals, individuals and classes; and the nature of relations, change, causation, mind, matter, space, and time.

Almost everything in metaphysics is controversial, and it is not surprising that there is little agreement among those who call themselves metaphysicians about what precisely it is that they are attempting to do.

1. Claims to tell us what really exists or what the real nature of things is
2. Claims to be fundamental and comprehensive in a way in which no individual science is
3. Claims to reach conclusions that are intellectually impregnable and, thus, posses a unique kind of certainty

REALITY ON METAPHYSICS

From the reflection that the surface of things often misrepresents what they really are, we must decide what is real as opposed to what they appear to be. There is a need to ultimately specify what different kinds of things there are in the world. What then is metaphysics, and what is a metaphysical argument? A metaphysician is concerned with advocating, articulating, and applying a set of basic interpretative, categorical principles to answering this question. These principles cannot be grounded in either conceptual considerations or an appeal to empirical fact.

METAPHYSICAL ARGUMENT?

There are no absolutely neutral data in the study of metaphysics. The metaphysician has the duty of explaining all the facts as he sees them; he also has the privilege of being able to decide what really counts as fact. A major point of metaphysics is that matters of fact have no relevance unless we solve them, and they cannot be solved without our thinking. Hence, the metaphysician is logical, analytical, speculative, subjective, and objective.

Many present metaphysicians are analytical philosophers largely influenced by logical positivism or ordinary-language philosophers. These philosophers do not questions more than the language itself—what inconsistencies lie in the language that prevent us from discovering what is real.

THE PURPOSE OF METAPHYSICS IN SPORT/MOVEMENT

- To review and analyze the history of the subject.
- To gain insights from the past and present.
- To make intelligible what is unintelligible.
- To make clear what is vague.
- To understand the whole and parts.
- To question, to seek, to find the nature of sport as we participate.

Some metaphysical questions about sports and movement are about real life experiences, are fundamental, and are uniquely self-critical:

1. Why did you choose this field?
   a. What are the ramifications of this question to your personal life?

2. What intrinsic benefits do you get from movement education, administration, coaching, and/or teaching?
   a. Can you apply what you learned from one class to another or to the real world?

2. What gifts do you have to give to this field of study?
3. Have you made a commitment to this field of study?

MOVEMENT PHILOSOPHY EXAMPLE: TO ALL OF THE OUT-OF-SHAPE PEOPLE IN THE GYM STARTING A NEW YEAR'S RESOLUTION

Before you start reading this particular example, I want to make sure that you understand that this example is not positive or professional in the movement world, but it is how some personal trainers feel about their clients and participants. Therefore, it is an example of this writer's personal movement philosophy.

> The gym in January, busy with wide eyed excitement and aspiration, is the exact spot where people swarm to begin their New Year's resolutions. Getting in shape is everyone's desire and is the number one reason why you will see multitudes of individuals in the gym early in the new year beginning to push themselves, lose weight, and become the best that they can be. You can see them harness all this drive and courage and actually do something by taking charge of their physical fitness.
>
> This greatly sought-after idea of starting anew at the beginning of a new year and using this drive to completely change one's physical fitness level and appearance, while seemingly a great thought, is juvenile and stupid. Yes, I am talking to all of you chocolate-covered chetto-consuming couch potatoes who, after seeing all the flashy gym membership advertisements around new years, decide that maybe you are just a few hundred belt sizes too large. And, let's not forget about you deathly looking twigs with two percent body fat, and even less muscle, who have the falsified idea that pumping a 10-pound weight with your biceps for a few days will turn you into a shredded body builder who has pencil-drawn indentations for abs. It is all of you measly excuses for fitness gurus who decide to crowd the gym and wipe your disgusting flab all over our precious equipment for a few weeks (and then drift back to the dark cave that you call a living room and continue to watch "The Biggest Loser" and dream of a day when your belt of shame, developed over many years of gourmet ice cream, will magically fall off) that drove me to express my concerns.
>
> I am not Arnold Schwarzenegger or Lou Ferrigno; neither am I that individual who lives and breaths in the gym. I am, however, a fit individual who invests a substantial amount of my own time and dedication into working out and striving for perfection in the gym. Having been going to the gym consistently year-round for several years now, I have been present during every extrinsic wave of out-of-shape slobs clogging up the gym and dampening the dedicated body builders' routines. But, of all of these gross dilemmas, none compared to the Winter of 2012 immediately following the return from Christmas

Break. Hordes of gross frat douches who could barely lift a weight and had zero experience in the gym flocked in their cut-off shirts, showing off every anorexic looking rib, as well as their non-existent chests, hoping to look good and impress the sorority girls, who similarly have invested more time and energy in their looks and appearance then they do in actually losing their cottage cheese thighs that they insist on showing off to the world.

The location of the first and most irritating facet of this enigma is in the parking lot (and adjacent parking lots). Upon arrival to a quaint and inviting gym in January, which frequent gym attendees normally park in front of 11 months out of the year, we were faced with a sea of parked cars that stretched out as far as the eye could see, along with a plethora of lazy six-pack seeking wusses who were limping to their cars after a 20-minute walk on a treadmill that they considered a good workout for the weeks. The worst part of this experience is watching in disgust as people will circle the front few rows closest to the entrance for 20 minutes waiting to find that desired parking spot with the shortest distance to travel. You are going to the gym to workout! All of you pathetic "fitness seekers" should be parking in the next town and walking to the gym in order to maybe burn off some of those exuberant jolly rolls on your jiggly corpuses.

I do not mean to offend individuals who are seeking self-improvement, nor do I seek to dampen the spirits of those who truly want to get in shape. I completely understand why you decided to begin a workout plan and incorporate fitness into your lives and how the extrinsic motivation of the gym is more beneficial attempt to work out and immediately quit are both de-motivating and frustrating to many others and me. Your parking is terrible, and you make it impossible during the month of January for us to workout, let alone move, around the blobs you create in an attempt to look like you know what you are doing.

My advice is to run around your block, do push-ups and sit-ups at home, purchase some kettle bells or dumbbells for your living room and lift those during the commercials of your precious TV shows. If, after all of that, you feel like you are actually motivated enough to start a real workout plan and stick with it, and you won't relapse and slip into a TV comma eating popcorn and ice cream from the container again; then, and only then, may you come to the gym, but no not park in my spot or get in my way!

Very respectfully,
A consistent gym attendee
(I. Isaac Wilson, personal communication, January, 15, 2012).

THINKING ACTIVITY

Read and observe this quote from Lance Armstrong's book *It's Not About the Bike: My Journey Back to Life*: "It's ironic, I used to ride my bike to make a living. Now I just want to live so that I can ride" (p. 119).

Imagine when cyclist Lance Armstrong started coughing one morning in the Fall of 1996; he initially wrote it off as just another after-effect of his strenuous training program. Then he saw the blood spattered in his bathroom sink, and at that moment, he was transformed from a tough professional athlete to a vulnerable man rushing to save his own life. He was coughing up blood, because testicular cancer, relatively common among men his age, had spread to his lungs and to his brain. The cancerous testicle was removed immediately, and he underwent brain surgery shortly thereafter. Aggressive chemotherapy was administered to shrink the tumors in his lungs. Almost overnight, the central question in Lance's life was not whether he could win the Tour de France, it wasn't even whether he'd return to the sport. The question at this point was: would Lance live or die?

Observe and analyze: Imagine Lance Armstrong's struggles by stripping them down to their most basic parts: a human being, biking, survival. Where does the meaning come from?

Observe and consider that, in 1999, Lance Armstrong did win the Tour de France, and the victory meant more to him and to others around the world than it seemed like a bike race ever could. The meaning came, first, from the fact that he had stared death in the face and survived the grueling treatments. But, it also grew out of some serious soul-searching about the purpose of his life and the values that would shape it. Finally, the victory had meaning for Lance—became a goal for Lance—precisely because it meant so much to others. It gave hope to thousands with cancer around the world (Reid, 2002, p. 119–120).

Question: Why is winning so important? Is winning still important when you are stripped down from Tour de France title due to performance-enhancing drug accusations?

CHAPTER TWO

Dualism in Movement Activity

APPLICATION OF PHILOSOPHY TO MOVEMENT ACTIVITY

Let us consider how we view training or preparing for physical activity. What do we do when we train? What sort of language do we use in getting ourselves "psyched" to train? What sort of language do we use when we talk about our bodies as we train? For generations and generations, we have spoken of the body as if it is a mechanical thing that is somehow attached to our minds and our spirits. This mechanical thing is something we deal with. For example, how often have you heard the above statements in training: Don't think, just do it? Train the body. No pain, no gain. This sort of language infers that there is something different in body than there is mind; that there is a separation of ourselves into two different parts—an objective thing—the body, and a subjective thing—the mind. We all do this language separation thing, and we have for centuries. Should we be concerned as physical-activity specialists? Well, of course, I am going to argue that we should, because how we use language in relation to our body has meaning. What sort of meaning? Some rather important meaning that will affect you throughout your professional life.

If there is a separation of the mind and body—there are two distinct parts—then one part must be more important than the other. Now here's a real chicken and egg philosophic sort of question—if they are distinct, then which is more important? Is the mind more important than the body? This is a slippery slope, because there is no answer—if you do not have one, you do not have the other ... well maybe, there are instances of individuals who have a great mind housed in a non-functioning body.

Can we have mind without body? Can we have body without mind? What is the relationship of mind to the body? Stephen Hawking gives me pause in how I am going to argue my case about mind/body. Hawking exists in a body that has deteriorated to the point that he has little gross motor ability. However, the body still does exist and when the body no longer exists, Hawking will no longer exist.

As a movement activity specialist who suggests the power and beauty of human movement, I propose that the body/mind is a unity and not a duality—meaning that we are a body/mind and not a body and mind. The majority of the world is in the duality/dualism mode. As I said earlier, there are numerous examples of how we actually practice this value-laden perspective of *mind* and *body*. We see the body as separate from the mind, "Don't think, just do it." We see the body as innately different than the mind, "Train the body."

If we see the body and mind as a duality, then in a hierarchal sense, one must be more important. Unfortunately, in our field, most bodily knowledge is not considered good, or as important as, mind knowledge. The reality is that the rest of the world sees bodily knowledge as inferior to mind knowledge. For example, "Chemists are smart; dancers are challenged." In the movement field, most of us body/mind seekers are always treated a little differently than the rest of the academic field. We are never thought to be as important as the rest of the university thinkers. Perhaps you have suffered some of these indignities yourself. For example, when you tell someone your major, and someone else snickers or makes some disparaging remark about your major, this implies that what you study is not very hard and is not very important compared to what they study.

How is this further played out? In the US, the general population sees bodily knowledge as secondary or of a lesser quality than mind knowledge. In No Child Left Behind no reference is made to movement, and some schools have begun to remove recess from the schooling experience, because more time is needed to learn.

How did we get dualism to begin with? We got it through history. History of the concept of dualism, the notion of the separation of the mind and body can be linked to Plato—of course, as Alfred North Whitehead said, "All western tradition is nothing more than a footnote to Plato" (Sharon Stoll, personal communication, January 21, 2009). And, it is further described by Rene Descartes, the 16th century philosopher, mathematician, scientist and writer who is dubbed the father of modern philosophy. He was a pretty important character who influenced philosophy and thinking and is still is doing so. Much of our perception of the thinking process today is influenced by his writing, and only in the last 10 years is neuroscience finding that Descartes was wrong about much of his ponderings. He is known for his basic concept: "Cogito, Ero Sum", or to practice my French, "Je pense donce Je Suis," I think therefore I am.

I think, therefore I am is a direct statement that thinking defines who we are. That there is a thinking being that makes the being exists. However, this separation does greatly affect our profession and our practice in movement activity. Most physical activity research views the body as an instrument to accomplish applied, proven principles. Movement activity participation, education, and research are extrinsically oriented to help the body jump higher, run faster, play more efficiently, learn more effectively, and so forth. This duality of thinking has had a direct effect on who we are—who we are dualists.

In fact, most of what we study and research in physical activity is centered in a belief that dualism works. For example, if we consider what you will learn in most of our scientific classes about physical activity, the knowledge base comes from what we know as "research." Sport research directs itself toward viewing the body as an instrument to accomplish applied, proven principles. But, movement philosophy class is moving away from this dualistic thinking into a more mind/body–value laden perspective. On the surface, dualism, as applied to movement activity, seems to promote a respectful scientific regard for the body through its main expressions of sport, dance, physical education, and recreation. Dualism appears to promote physical activity participation for personal performance, fitness, and competition. And, the sort of education that we use to reach the goals listed about becomes technical, systematic, and methodological.

Dualistic thinking of a separated mind and body appears to make us seem very important. We get people fit. We do so in a very systematic fashion. We do so through scientific principles. That's who we are. Yes, this dualistic way of thinking has been who we are from the very beginning. Even though science is who we are, science and how we do business has not convinced the masses. Why not try something different? Why not think differently about this thing that we do—which it is not outside of us, not an addition to who we are, but which is the measure of who we are.

Dr. R. Scott Kretchmar wrote Practical Philosophy of Sport and Physical Activity (2005) and argued that every time you lift a weight, you affect the whole man, not just a muscle. Kretchmar also posed some intriguing questions:

- Are there any good psychological or philosophical reasons for separating persons from their bodies?
- Is it good to treat bodies simply as machines?
- Are minds more important than bodies?
- Are "intellectual" professions more important than movement-based professions?
- Does physical education have a stake in how persons are related to their bodies?
- What difference does it make if minds and bodies are separate and independent parts of persons or if minds are more valuable than bodies?

In addition, Kretchmar (2005) has crafted five different types of dualism: substance dualism (p. 50), value dualism (p. 51), behavior dualism (p. 51), language dualism (p. 51–52), and knowledge dualism (p. 52).

SUBSTANCE DUALISM

Substance dualism is based on the notion that physical matter and thinking are completely different from one another. The body can be measured, dissected, and treated like any other object. The body is nothing more than a moving machine whose actions can be described in terms of levers, forces, and material properties. Treating the body as a machine has resulted in better health, better movement performance, and far better understanding of how everything from genes to human physiology influences us.

VALUE DUALISM

Plato placed mind above body, preferring thoughts, ideas, and perfection over emotions.

BEHAVIOR DUALISM

According to behavior dualism, all actions are composed of two parts, thinking followed by doing. The body cannot act on its own, because it is only a machine, so it must await commands from the mind.

LANGUAGE DUALISM

Language in prose, poetry, and other spoken and written forms, as well as mathematical symbols in physics and other sciences, are commonly thought of as intellectual in nature. IQ, SAT, GRE and other intelligence and academic achievement tests are thought to show the workings of the human mind at its best. Symbol systems that do not rely primarily on words or numbers (music, painting, and sport) are often associated with nonintellectual endeavors, and they enjoy lesser academic status.

KNOWLEDGE DUALISM

What we know? Why some sport techniques work better than others, or how and why circulation improves with exercise interventions. Often times we wonder how we can perform certain skills without being able to explain them. Knowledge dualism argues that people with this kind of knowledge do not really understand what they are doing.

THINKING ACTIVITY

Consider the following questions:

1. Is it not correct that people have both minds and bodies?
2. Are minds more essential than bodies?
3. Do our minds command our bodies?
4. When we train, coach, teach, instruct our students/clients/athletes, should we only instruct their minds?
5. Is kinesiology a divided field with two research plans and two kinds of interventions (one for the body and one for the mind)?
6. To understand the movement or to improve sports performance, do we have to deal with two different things: cardiovascular system and oxygen utilization and stories, emotions, and experiences?
7. When you train your clients or coach your athletes, whom do you train/coach? A machine? A mind?

CHAPTER THREE

Wholism in Movement Activity

How can we get rid of this dualistic way of doing exercise, play, and physical activity? To possibly try to solve this issue, we can try to think about movement subjectively and not objectively. What is the difference? Consider this example: look at your hand—how many fingers are there? Describe your hand: Did you describe it by color? By length? By size? By texture? If you did, you are thinking in an objective fashion. Objective thinking is usually tied to material aspects, such as numbers, size, length, shape … something that can be measured in an objective fashion. Now, picture two hands holding one another. What objective features do you see? Two different genders? Two hands physically linked? Consider the meaning of the two hands as they appear together. These two individuals are holding hands—they are touching and being touched. What is the meaning of this action—it is a subjective knowing—it is … what? Love? Sure, there is an objective bodily response to loving another, but the subjective is the power undercurrent of these hands together. Can we measure love? Can we place love into a formula? No—it is a very subjective, meaningful, individual sort of knowing. For example, the term embodiment is often used to explain that we are our bodies and that our bodies count for something. Hence, they are not just objects to be manipulated and measured, but they are a center of all our experiences.

I am challenging you to try thinking subjectively … this may not be easy, because science does not like the subjective, but suppose we take a leap of faith and think about movement subjectively, rather than objectively. In relation to dualism, objective thought carries the day. That is, dualism is the belief that we are an objective body and an objective mind. In contrast, a subjective view might be that there is not two different objectives but a subjective connection between the two.

One of my favorite movement philosophers is R. Scott Kretchmar. He is an active lover of movement. He is a lifelong runner—a passionate supporter of everything that has to do with movement. Dr. Kretchmar has written much about the dilemma of the mind/body dualistic approach and argues for a more subjective approach to the view of the relationship of the mind and body. He explains:

> My mind and body are not two but one, they work synergistically. Mind and body support one another for physical and mental purposes. They are more powerful when realized as co-existing and working together. Mind and body do not seem to act on another externally, as indeed they would if they were radically distinct entities. It is not accurate to say that a freestanding, independent mind tells the body what to do, or that a freestanding, independent body responds that it will or will not obey. Rather, when individuals think of purposes (like kicking goals in soccer) and supposedly tell their bodies what to do, they already are their bodies.
>
> (Kretchmar, R.S., 1993, p. 38)

Notice how often we objectify the body, such as referring to the body in impersonal terms like "it," Tending the body, training the body, saying phrases such as "get the body fit," and "no pain, no gain." In these cases, body is separated, and it has to be trained like you would train a dog. Avoid referring to the body as something to bring along with you or something that is "not working" or saying "my head is not in it."

Try to not pretend that you can manipulate bodies—work on them, train them, educate them—without affecting the whole person ... Heart rates do not get produced in a vacuum. They come with pain, accomplishment, interest, boredom, love, hope, and virtually any other affect that can be thought of. They come attached to human purposes and embedded in human stories. They are related to what people have been, what they are now, and what they hope to become (Kretchmar, 1993, p.41).

If Dr. Kretchmar is correct, perhaps we need to spend more time thinking about strange and wondrous things like the subjective, rather than the objective ... What would happen if we did more subjective thinking? Suppose we take some of Dr. Kretchmar's philosophy and put it into practice and see, value, and love exercise and movement as the most important things we do. What would happen if we could not sit at a computer, but had to walk to make the bloody thing work? And, what if we exercised instead of sitting when listening to a lecture ... because what we are doing now is some sort of dualistic philosophy ... so *on your feet* and do 20 jumping jacks, or 20 tour jetes, or 20 somethings ...

How can we get to the point of movement like the Ancient Greeks. The Greeks, beginning with the first great Western Tradition Thinker, saw movement and exercise as a part of being human, civilized, and educated. Socrates did not write anything, but left it to his student Plato to let the world know of him and what he believed. Socrates taught his students while walking and talking. In fact, Socrates liked to do his discourse at the Gymnasium … And the gymnasium was the most important place for social gathering and learning. What if the most important room in every school was the gymnasium, where students learned all of the content including math, science, reading, language … what if the gymnasium was the beginning point, rather than the "if we have time" point?

If the gymnasium was the most important place, and all learning was motor learning, then why not think differently about physical activity this thing that we do. After all, isn't movement the measure of who we are? Dr. Kretchmar is pushing us to think not objectively but subjectively … to move away from the objective nature of dualism and move forward to a more wholistic approach.

I mentioned earlier that all learning is *motor learning* … all learning demands the use of motor neurons and perceptive skills to learn any task. We, in movement activity, are about gross motor skills, but fine motor and perceptual skills are also a necessity. We are the motor control experts. We experts in motor learning and motor control should also be important people in any field of endeavor, and especially in schools. And, because we are motor beings—human beings who are made to move—we should be moving totally and wholly in everything that we do. We learn by moving!

We are movers and that we should be passionately connected to movement in many ways. It is not about the mechanics of movement, the health rewards of movement, or the social benefits of group participation. What if we use a different sort of language about movement … in which we enjoy what we do, rather than language that sounds boring, painful, and gruesome.

Dr. Kretchmar argues that physical activity should have a *wholistic purpose*. The purpose should be about helping develop and maintain a good life. The Good Life refers to an overall life condition and set of experiences that we regard as desirable. While most people aim at good living, there is considerable disagreement about what that exactly is. There are probably hundreds of ways to achieve something called the good life, and there are undoubtedly many patterns that are comparably good (Kretchmar, 1993, p.111).

The good life is composed of experiences that are appreciated for their own sake. The good life must be meaningful or have purpose. One event must be meaningful or have purpose. One event must be connected to another, and these events must be headed somewhere important as in the development of a story line. Survival and long life, by themselves do little to assure good living. A good life should entail pleasure and fun, which is almost universal. Movement should bring meaning to your life. The good life should be meaningful, and physical activity should bring meaning (Kretchmar, 1993, p. 224).

If it is about enjoyment, and about a life story or narrative, we wouldn't be "working out", we would be "playing out;" it then becomes something I look forward to throughout my day and not the drudgery of "I have to go workout!" When I speak of drudgery of exercise, I think of something that I have to do, because it is good for me, like eating vegetables. Unfortunately, most folks do not seem to be motivated by "what is good for them." What is good for us is usually called "prudential," meaning doing something, because it shows good judgment to do it, or using common sense in doing something. If vegetables are good for me, then it is common sense to eat vegetables. In physical activity, we have used this prudential way of thinking forever in motivating folks to exercise. It is good for us; science and health shows us how good it is, so you should move, because it shows good judgment and common sense to do so. Unfortunately, this prudential way of thinking has not motivated people to move. If we follow Kretchmar's advice, we should think about movement as something more valuable—something tied to value of our lives, our rich experiences in life—not the experiences of drudgery.

Meaning is individual. How about you? Why do you pursue physical activity. Is it for skill, fitness, knowledge, or pleasure? Even though you may do movement for prudential reasons—fitness and knowledge—a better way to view movement is as developing skill; usually if we are more skilled in physical activity, whether it is biking, skating, dancing, or running, we like to do the activity because we enjoy it. The joy of playing and the joy of moving should be ranked above fitness and knowledge to help us experience the good life. Again, what is the good life? The good life is presented to us through lived and learned experiences.

Dr. Kretchmar has introduced us to the concept of life narrative—a life narrative is your story—your story of movement, your story of physical activity, your story of a body/self as a living, moving being. That is what we should be. That is what we should practice. That is what and who we are. We are life movers or we should be!

Try to begin describing yourself and thinking, "I am a mover, a runner, a biker. I no longer see myself as a slave to exercise. Exercise/movement is who I am." When you define yourself by movement or activity, it becomes a joy – not work. Each and every movement activity should be regarded as special. The values of fine movement are so great a prize that everyone should want to achieve them and help their neighbor achieve them. Think "I like movement and I like me when I play. Dr. Kretchmar says:

> Discovery. Human movement is a dialogue between persons and a spatiotemporal world. The dialogue is given life by purposes—to play, to win, to score, to kick,to show. As the dialogue unfolds, discoveries typically trip along one after another. People learn about themselves—their personalities, their capabilities, their intensity, their determination, their generosity, their fears, their tenderness, their prickliness, their capacity for love, their potential for hate. This information does not come inscribed on parchment. It comes as human beings jump, through their victories and defeats, when they swing or pirouette, as they fall or dive, while they pass to a teammate or get shut out of an offensive scheme. This process of discovering can be valued for its own sake.
>
> (Kretchmar, 1993, p. 195)

Satisfactory experiences that build a coherent and meaningful life narrative take precedence over those that are isolated moments of pleasure; Developing exercise life stories and realizing a body/self and defining your life through exercise/movement can count for satisfactory experiences. Exercise is not something to be worked into the day. Exercise is a joy to look forward to. In describing yourself, terms are used like: "I get to exercise today," not "I have to exercise today." "My day begins with movement." "I am a dancer, not I dance." "I am a mover, not I exercise." "I dream about movement." "I look forward to being creative in my movement. I do not worry about parking close, or avoiding the stairs. Instead, I incorporate movement;" "I add movement, I take the stairs; I park further out; I enjoy the challenge, the experience, the activity of movement." Perhaps the most fundamental and traditional experience of a coherent life comes with developing and living a story. Surround yourself with items that say you are a mover. Do not be afraid to keep your exercise shoes right by the door, your water bottle always at hand, your favorite fitness magazine nearby, or other possibilities of items in our marvelous field of physical activity. Pleasure of movement is often associated with play that is intriguing, gripping, fascinating, or captivating, one that is enjoyed for its own sake. Movement has a pronounced and powerful aesthetic component. Aesthetic meaning pleasing to the senses: sight, sound, smell, touch, taste, and proprioceptors. For example, when I run, my eyes provide the beauty of the river, the trees, the sky … When I run, I hear the rhythmic pattern of my own feet, my breathing, and the beat of my own heart. When I run, I smell the pine trees and the river; in spring, I smell the blossoms around me. Even if I run in the city and smell car exhausts and wet pavement, it brings me joy. When I run, I taste the salt of my own sweat, feel the cold air, and even almost taste the smell of pine trees. When I run, I have the joy of feeling spent, and my joints and muscles give me the pleasure of the aesthetic experience of fatigue. It is the effortless attempt that I experience (Karen Rickel, personal communication, June 12, 2010). Dr. Kretchmar agrees:

> Pleasure often occurs in environments where something of a spell has been cast over its participants they are give to the experience—so given, in fact, that they are not sure why they spent so much time there. When asked after the fact why they gave so much energy to dancing, playing field hockey, or riding a bicycle, for instance, these players will often say simply that they experienced a great deal of pleasure. Or more likely, they will just say it was fun.
>
> (Kretchmar, 1993, p. 168)

You already have a good start; that is why you have chosen to major in kinesiology. You are a dancer, an athlete, a runner, a re-creator, a player … it is time we celebrate and see ourselves as movers, and when we can scream to the world, "I am a mover," then we are ready to motivate and inspire others to be movers, also. In your narrative of life have you considered what the purpose of your moving is? What should it be? I am talking about the things that we do that are meaningful to us just because they exist: a beautiful ski run; a dance that makes you want to dance more; putting on your skates; smelling the ice, the field, the court—and it does have smells—the smile of a friend; holding the hand of your loved one; the giggle of a child; racing with grandpa in his wheelchair! Movement should bring meaning to your life.

What if we stopped thinking of success as something that is mental, instead we thought of movement as positive perspectives of "I play," "I move," "I dance," rather than in how many reps I did not do or what I did not accomplish? What if we saw movement/play as so satisfactory that we build a meaningful life narrative—a life story that is told through who we are as movers—that we are not satisfied with bits of movement, but that our life is movement. What if we saw play as something we *have* to do, that we *get* to do, that we *cannot wait* to do?

If the distinction between scientific knowing and bodily knowing is not significant, then we would be like all other disciplines. We do what all other disciplines do, help people communicate, explore, invent, create, figure, produce, and compute. We do so through gesture, waving, running, jumping, walking and so forth. And, that is the point—we need to see movement as the necessity of who we are, rather than the tag on to the more important things of life: work! We would be liberated to communicate and produce in many different wholistic ways, and not just with disembodied hands or words or numbers. Because we would not be auxiliary to the real world, we could liberate the human from the bonds of mind/body dualism. So, the answer lies in you and me. We need to think about our body/self in a liberating fashion. We need to think and speak about our mind/body as a moving being, and our moving mind/body develops a life narrative. I am a *mover*! I love to *play*! I love to *dance*. I love to *compete*.

When you are able to make this jump about who you and are PROUD that you are a MOVER, then we can actually start the journey to help others in their life narrative growth!

THEORETICAL AND PRACTICAL PROBLEMS WITH MOVEMENT DUALISM

Substance Dualism

Treating the body as machine has resulted in better health, better movement performance, and a far better understanding of how everything from genes to physiology influences us. Dualists have never explained how mind and body affect one another. Do we ever have pure mind and pure body? If mind and body are separable, how much attention should we give to each? Do they each deserve equal amounts? Perhaps one deserves 70% and the other 30%. As kinesiologists, once we divide the person into parts we run the risk of becoming a mere "body shop." (p. 53–54).

VALUE DUALISM

Our passions could lead us astray and our mortal bodies would decay (Plato, 1951). John Wooden is an inspiring example of mind over matter. It is not clear that the mind is the source of our salvation and the body the source of our ills. It is one thing for John Wooden to guide a group of highly motivated and skilled individuals toward perfection, but what about average athletes, students, and clients? Should we agree that intellectual education is more valuable than physical? Should we require undergraduate kinesiology students to take mostly theory courses with little or no performance (p. 54)?

BEHAVIOR DUALISM

When athletes get into "the zone," ideas and actions merge into one, and behavior flows effortlessly. Should our profession emphasize knowing the theory of moving well or developing the skills to move well? Because we have precious little time with our students and clients, efficiency might suggest that we focus on theory—once they understand the theory, they can work it out on their own. When athletes are in the zone, they do not reach first and act second (p. 56).

LANGUAGE DUALISM

Movement can become very personal, expressing who we are and what we value all without a word or mathematical formula. Does language dualism imply that we have not done our job if we have not encouraged our students and clients to convert movement experiences to words (p. 58)?

KNOWLEDGE DUALISM

Is it possible to understand something without being able to explain it (p. 60)?

THINKING ACTIVITY

The purpose of this thinking activity is to begin a journey—a journey of the mind and body. We learned the meaning of dualism and wholism, integrating a healthy, active lifestyle and focusing on the nature and purpose of a healthy, active lifestyle. I am asking you to respond to a following question: **While participating in sports or exercising, how do you think of your mind and body—as separate entities or working together?**

And if so, how is this manifested? Keep in mind that there are multiple ways to respond to this question.

REFLECTION ON MIND/BODY DUALISM AND WHOLISM IN MOVEMENT ACTIVITY

While performing any sort of purposeful physical activity, I definitely identify my mind and body as a coupling that indeed works together in harmony. Evidence of this occurrence is apparent to me nearly every time I workout. I cannot have a good workout—meaning I am in a good mood during my workout and feeling even better after my exercise—if one of the components (mind or body) is not in sync with the other. Often times, I use gym time to reinstate this union of mind and body. More explicitly, I may have become so locked up in my head (stress from classes, relationships, etc.) that I become disconnected from my physical existence. Or, on the contrary, I may have been sitting so long (studying, in class, etc.) that my body is so stiff and suffocated that it distracts my mind and robs me of my ability to accomplish necessary mental tasks.

Perhaps it was my upbringing as an athlete—specifically as a gymnast—that bore this wholism principle so deeply within me. As a Junior Olympic Competitive Gymnast for 11 years of my life, many of the sport's unique disciplines found their way into the fiber of my being. Recalling, for example, that there was very limited to no talking allowed during practice, one's focus would remain entirely on executing motor movements with incredible precision, instead of potentially being distracted by an intrapersonal conflict between teammates. Taking this concept and applying it to the process of acquiring a new skill perfectly illustrates the innate mind–body connection. As the acquisition of skills was imperative to one's advancement in the levels of artistic gymnastics, this was a process that I had vast experience with. First acquiring a skill would require many repetitions of "drills," or repetitive physical conditioning of one part, section, or aspect of the desired skill (e.g., learning a Yurchanko vault required cart-through drills to condition driving of the back leg, blocking drills to condition the timing and strength of a strong shoulder push, and mat vault drills to simulate the back handspring part of the vault without the speed). Using the drill technique trains functional strength for the skill and body awareness to keep you safe during its execution, and perhaps most importantly, drilling motions takes the fear out of the skill so that when one does finally "go for it"—with all of the drilled parts as a whole, at speed, and with just the bare equipment—hesitations, air loss (getting lost in the air/no body awareness), and bailouts can be minimized, an accomplishment that yields confidence to attempt the skill again (and again and again and again) in order to fully master it. With this sort of mechanism being an integral part of my upbringing, it is a natural progression for it to translate into other areas of my life and eventually evolve into an extremely significant part of my very existence. I lack the capacity, much less the desire, to operate in a

Dualistic sense. Experiencing over a decade of training that began at a very young age, wholism is, indeed, just as much a part of my fiber as my family traditions and concepts of right and wrong. It is as plain as the notion of segregation that induces such discomfort for me. The stark line that dualism draws between one's mind and physical being seems almost discriminatory to me; being that discrimination is a behavior that I was raised to never passively abide nor take part in, segregating one arena of my existence "For My Mind" and another "For My Body" is inhumane and utterly intolerable. As such a rich blending of my mind and body has always been a fixture in my life, I do not believe that I could completely separate the two, even if I tried. Even though I am no longer a gymnast, I still strive to nurture the union (between mind and body) that it instilled in me so long ago. Particularly through my hot yoga practice and regular exercise regimen, I ensure that the sanctity of that bond (between my body and mind) is never shaken.

However—on a slightly darker note—I do admit that I did suffer a period of dualism in my life after I was raped by an ex-teammate. (I used to row for the University of Oregon. He was one of the varsity men.) The schism happened during the act itself—and I can still recall to this day—I experienced the episode as a body and as a mind. I described it to my counselor as if I watched it happen, like I was no longer present in my body and was watching passively as someone defaced it. It took two years of counseling and a bout of depression to finally settle me back into my own skin, and mend the severed bonds between my Body and Mind.

Envision an affair between two intimate, exclusive lovers if you will. My Mind spent years being jaded by a Body that betrayed it—not dissimilar to a partner who was devastated by a cheating spouse. Trust had to be rebuilt and respect for one another had to be regained if there was any hope of salvaging their future together. Counseling was involved to tend old wounds and elicit long lost communication practices, but bit by bit, they (my Body and Mind—the lovers) found each other again.

Happily reunited, My Mind and Body are looking forward to a lifetime of togetherness. Sworn by love, tested by violence, they will never grow apart (Molly Garner, personal communication, March 12, 2013).

SUMMARY

The challenges we face in movement and physical activity are just as mental as they are physical. What if we tried to balance all three major components of the theory? A unity must exist between the mind, body, and spirit to ensure optimal and balanced improvement. From the physical aspect, proper program design is essential. Every exercise has a prescribed amount of repetitions and sets to optimize desired results. Every rest interval is important to work the correct physiological systems within the body. Every workout is written down before hand and every weight specifically chosen to pattern proper periodization. But sometimes, things just do not go according to plan, so we have to improvise. The gym is very crowded sometimes, and we do not always get to do the workout written down. We have to improvise on the fly and implement a different exercise that resembles the desired movement pattern and stimulation. This balanced outlook helps with continuing to train at the highest level without sticking to an unrealistic training program. From the mental standpoint, we have to stay focused. Every warm-up is just as much mental preparation as it is physical. We prepare our body for the upcoming intentional work. One must induce a certain amount of stimulus upon the body to cause a response of growth. The closer you are to your genetic potential, the harder it is to create this stimulus. We have to mentally prepare our bodies for a serious challenge to make sure for the best effort. We have to have a positive and assertive attitude that is built upon diligence and commitment. We have to be very disciplined outside of the gym, too. We have to maintain a high level of responsibility to ensure that we are consuming the proper types and amounts of foods to promote optimal recovery. These mental components are very important and cannot be overlooked. We also have to maintain a healthy spiritual outlook to reduce stress and optimize overall life quality. Spirituality does not have to involve religion; it can just be your connection to and awareness of your environment. Close your eyes and relax for 15 minutes a day while you observe your surroundings and let your mind peacefully wander. If there was only one benefit you took away from this activity, it would be reduced levels of stress. These lowered stress levels would physically benefit you by balancing your hormones and reducing the production of cortisol. Even this small benefit can improve recovery rates immensely.

Can we strengthen character through movement, just like muscles? The more you confront resistance, the more you struggle, and the more obstacles you overcome, the stronger character you build. Out of the gym, can we see the results in all areas of our life? We could be humble but hungry, peaceful but assertive, and most important, happy. You become what you think about and what you practice.

CHAPTER FOUR

Passion for Movement Activity

The true beauty of what we do in movement activities can be hidden in our metaphysical knowledge and experience. As athletes, movers, and exercisers, we may be passionate about this subjective part and live in our sport or physical activity experiences. We may also consider that we belong within the special group that receives the emotional and physical gifts that come from immersion and participation in movement, exercise, or organized sports. The opportunity to run, lift, and play games may be exuberant to some individuals.

Some of us can enjoy the pure fun within basic movement and activities that bring out the best in us when we move, participate in sports, play, and exercise. The most basic pleasure associated with any activity can be called the "joy of movement," which derives from tactile and somatic sensations (Todd, 1979). In fact, many people like to play sports, exercise, and move because they like how their body feels while conducting those activities. For the rest, movement, physical activity, or exercise may appear to be chores. Todd (1979) explains that a swimmer may feel an additional beauty and an extraordinary sensations being in the water or feel powerful and strong as he/she pulls him or herself through the water.

In dedicating their performance to the gods, the Greeks may have been giving religious expression to an aesthetic truth. The beauty can take several forms (a beautiful feeling in meeting a challenge, in pushing one's limitation outward, in going faster, harder, or longer than ever done before). There is beauty in feeling one's mind and body working together at full capacity in complete harmony.

Can sport/movement activity come just with toughness? But, compassion, kindness, and love. We may have been a lot happier and less confused sticking to good, hard, and quantifiable data like VO2 max and training zones, but how do we make people love the activity or make some athletes love their sport? How can we flip the internal switch that changes us into the passionate movers? I remember being passionate about physical activity that my parents had to yell at me to slow down. when my parents had to yell at me to slow down. Every game I played, I played at top speed, at 100%, making it last time in my life I'd ever be hassled for going to hard and too fast. There may be some kind of a connection between the capacity to love and the capacity to love sport and exercise. Both the capacity to love and the capacity to love sport and exercise can depend on our desires; passions; appreciation for what we have, rather than what we want; and patience and understanding.

Maybe being better in the capacity to love can make us better in the capacity to love and appreciate the movement activity. Phil Jackson, one of the greatest basketball coaches of all times, summarized it best.

> You guys need to get together and remember what you're doing this for. You're not doing for money. It may seem that way, but that's just an external reward. You're doing it for the internal rewards. You're doing it for each other and the love of the game.
>
> (Jackson, 1995, p. 162)

MOST AMAZING PHYSICAL ACTIVITY EXPERIENCE

I would expect myself to illustrate the process by which extreme perseverance was demonstrated, so that a necessary gymnastics skill called a back handspring could be acquired, but despite how exciting that moment was, it is far from the cornerstone of my physical activity experiences. Perhaps I tell that story, because it is one that I have told before—but that story no longer speaks to me. A simpler occasion resonates with me instead: conditioning. Strange right? I think so too, but it could not be more significant to me. The meaning in this seemingly simple activity is a result of the oneness that I felt within my own self, as well as the undeniable unity that was shared between teammates as we took on the challenge of Coach Ferra's Level Six summer conditioning regimen. Gymnastics is, to many people, an "individualistic" sport, where the emphasis is placed on the performance of an individual, and "teams" are merely collections of athletes who happen to practice together, nothing more. This perception is correct to a certain extent but really fails to acknowledge that the strength of that lone person performing on an apparatus before stone-faced judges emanates not from her, but from her teammates behind her. I value that 2004 summer at Gymnastics East more than any single physical performance simply because it was shared survival, a test that we—myself and 15 of my teammates—endured and were successful in. During the summer, our practice schedule went from 4 hours, four times per week, to four 5-hour and one 4-hour practice per week. The gym had no air conditioning and was stranded in an island of sizzling hot blacktop (i.e., an industrial park). Saying it was "hot" in there really does not do it justice. Practices in the summer consisted of 30-minute to 1-hour sessions on the four events, and another hour dedicated exclusively to conditioning. That hour was my heaven and my hell. That hour was dreaded by all and regretted by none. That hour was what set us apart and kept us together. In that hour of snap-downs, V-ups, chin ups, sprints, pit jumps/crawls, donkey kicks, killer crunches, standing tucks, suicides, over-unders, P-bar handstand pushups, and hollow holds we lived and died together. We turned our minds off—because they were screaming, "You've had it. There isn't any more left to give. This is your breaking point. Stop!"—and turned on our bodies and hearts to find strength (physically and psychologically). I treasure that summer of staggering heat, muscle failure, and crippling soreness, because of the sanctuary that it built in and between all of us.

This oneness cannot be found elsewhere or replicated, save perhaps those who serve in the active military. The simple fact that we survived—that no one was, or would ever be, left behind—held us together as a group. Also, from that same fact, one concludes that the sanctity of her body–mind–spirit connection was one forged by fire and will never fail her. This was the most awesome physical activity experience of my life—bar none.

THINKING ACTIVITY

What was the most "awesome" physical activity experience of your life—an experience that you will always remember? Give the details, including year, activity, location, and reason why was it so important for you.

PERSONAL REFLECTION ON RUNNING

The clock ticks 2:40, signaling the end of a long day of classes. I can't help but think is, "finally." I race home, charge up four flights of stairs to my room, throw my backpack down at the foot of my bed, and begin changing. I am on a mission. I happily trade my True Religion Jeans for a comfy pair of DRI-fits, my lacy pink push-up for a sleek training bra, and my Old Navy flip-flops for the most precious piece of clothing I own—my custom made Nike Frees. My hair is ponytailed; the iPod is charged and ready; the Rec is calling my name. I am pumped! I practically prance through the doors of the Student Recreation Center, or what I like to think of as my second home in Pullman. It is at this point where I come to realize there is no distance too far. There is no limit to how fast I will run today or how long my body will last before calling it a day. Pain? What pain? There is no such thing as pain—only exhaustion. I will gladly tell the treadmill when I have had enough, thank you very much. Out of the corner of my eye, I spot my favorite machine, sitting idle and untouched. Jackpot! I make my move, place my sweat towel in a handy position, adjust my headphones, and start that belt. Oh yes! It is go time!

Running is not my sport. Whether it is indoor or outdoor, running is my passion. The sound my feet make when each one thumps on a tight treadmill deck or crunches beneath a patch of loose gravel; the alignment each stride makes with the tempo of an upbeat hip hop song; the immense sense of accomplishment that extra half-mile brings after mentally begging my legs to push through the pain—all of these are things I think about, long for, and live for. There's no greater feeling in my mind than bearing witness to myself become stronger, faster, and longer-lasting. Racing down a long stretch of empty road, winding through an uneven forest trail, or pushing my limits on a pristine treadmill is my ultimate getaway—all I need to pack is my iPod. While the thought of sweat pouring down their freshly powdered face may sound horrific to many girls my age, I somehow feel most striking when I'm dripping sweat. Sweat not only filters my body of all impurities, stresses, and anxieties thriving underneath my skin, but it is also a sure fire indication of an exceptional work out. Look at me. I went hard. Why do I run? Why do I breathe? It's a question that seems rather silly. In my mind, to run means to prove, to challenge, to defy; I absolutely could not imagine my life without it.

I run to prove myself, but not necessarily to others. I am not going to lie, I do feel superior when my machine is set for a faster pace than the person next to me, but for the most part, I run to satisfy a deeper fulfillment. If it makes any sense, I run to prove myself to myself. In my opinion, I have never been phenomenal at anything. I have never been the straight-A student. I do not know what it feels like to carry the team nor be that irreplaceable playmaker. By no means do I find myself unusually gorgeous; I look in the mirror some mornings and cannot help but feel like a 5-foot 7-inch pile of flaws and imperfections. Whether it is with schoolwork, looks, or even other competitive sports, I feel as if I have always been nothing more than slightly above average, despite my best efforts. Running has always been that one thing I know I am exceptional at. And maybe that is because, with running, or at least non-competitive running, there is a standard of excellence to achieve while competing against oneself.

Running truly is just as much of a mental work out as it is a physical one. I believe that as long as I am running, my body is always competing with itself—my head versus my legs, my heart versus my lung capacity. I know I am not always capable of winning this race, some days the shin splints or diet coke–induced side ache may get the best of me. But rest assured, the days I perform are the days I feel exceptional, victorious, and elite. To me, the best days are the days I surpass the expectations I set for myself on that treadmill, track, or trail. I run a little faster or a little further than I did the day before, and boom. No longer am I average. In my own mind, I am a legend.

CHAPTER FIVE

Movement and Meaning

Many individuals do not like to compete or be a part of physical activity. I suggest that you are not one of these people, because you *love* to move. You have chosen a field that is about movement, play, sport, and physical activity. You are truly lucky. But unfortunately, you are in the minority. Today, according to the Center for Disease Control and Prevention (CDC), the majority of the population is not moving at the level that they need to to burn enough calories to keep themselves from becoming unhealthy. The CDC is predicting that the majority of the population will be obese by 2020, if we keep up the present pace. We will likely not be one of the majority in 2020 who is obese and unhealthy. Why? Somewhere along the way, we have discovered the importance of movement, and somewhere, we have discovered that movement is very special, whether the movement is sport, play, exercise, or dance.

Movement philosophy is thinking about why we love to move. If we could take the why of movement and physical activity that we enjoy so much and give it to others, we would not be faced with the existing tragic statistical data.

Throughout this book, you will experience some wondering about yourself. For example, why are you here? I hope the answer is more than "to get a job." I hope the answer has to do with your love of movement—the joy you get from movement and that you hope to give that love to others. It is ironic how science has given us the majority information about exercise and health, but has failed so miserably to get people involved in movement. The majority of our population is not burning enough calories to counteract the amount of calories that they eat. As a kinesiology major, you will learn about energy expenditure and energy intake. It does not take a rocket scientist to know that if you eat too much and do not exercise enough, you will gain weight.

During my master's and doctorate studies at the University of Idaho, I was introduced to the fitness craze and jumped on the bandwagon as a fitness instructor and personal trainer who was going to change the world. After all, we had the science to motivate the world to pursue health and fitness. One of my professors was Dr. Damon Burton, one of the authorities in sport psychology and motivation, who gave me a plethora of information on how to motivate individuals to be physically active. I thought the world was going to be a better place because of my hard work, positive energy, and enthusiasm. The scientific knowledge would motivate people to move and be fit. There was going to be a revolution; I was there, and I was a part of it all. Science and exercise would change the world. It was not until I met my major professor, Dr. Sharon Stoll, who is one of the 100 most influential sport people in the world, that I realized that science has not trickled down to its participants, and we still do not understand why some people choose not to move and be physically active.

I still do not understand how the human body subjects did not get the job done. We have never had an epidemic like this that we have been able to track so thoroughly and see so clearly. For the past 60 years, we have had a dramatic increase in obesity, and there has been a 61 percent increase in diabetes since 1991. I can safely say that we are an overweight nation, moving quickly to obesity, and we appear to feel the need to eat more and more. We have the science to combat this problem. We are fabulously rich people. We are highly educated. We know what to do. The formula is clear: energy in equals energy out. But, we cannot motivate the population to practice good eating habits and commit to exercise.

It is quite simple—whenever someone forces us to do certain things, we want to know why we should do them. In movement, in particular, we want to know why we should move, what purpose it will serve, what good it will do (Kretchmar, 1993). Often we work with sedentary individuals who already embody their sedentary habits, who may be comfortable in their style of living, and who want to know why they should change now (Kretchmar, 1993). I was introduced to Dr. R. Scott Kretchmar when I first started my doctoral program. His book *Practical Philosophy of Sport* (1993) helped me realize a great deal of discoveries about movement and meaning. In fact, it made me realize that not all meanings are created equal, and often times, we can tell the difference. For example, we may exercise for some time and find the activity difficult and repetitive. The movement keeps its distance and hold on to its secrets. But, for some of us, the participation in such activities begins to take on a different shape. We can express ourselves with movement. When we establish this kind of intimacy with an activity, and we are frequently carried away by the meanings that come our way, reasons for continuing to participate usually no longer matter. We are hooked; we need to be involved; it is difficult to imagine our lives without this good friend (Kretchmar, 1994).

Maybe the solution and the answer lies in how we view the body. This drawing, by Toles, is a literal translation of Socrates: "Know thyself," or "An examined life is not worth living." Socrates' definition translates better to our world of movement. Who are we? Who am I as I move? Our commitment to science and exercise has missed a point, a philosophic point that was made centuries ago in which science has argued that we are two separate entities—a mind and a body. But, are we a separation? Are we a duality? Perhaps, we are more than just mind and body.

CHAPTER SIX

Ethics in the Kinesiology (Movement) Profession

In this chapter, you will learn about different moral and non-moral values and how they shape our profession. The focus of this chapter will be about the effect of non-moral values on your moral decision making, whether it is in competition, in the classroom, or in the gym.

NON-MORAL VALUES

Non-moral values are relative, which means that they are learned, social, cultural, and usually about the worth we place on extrinsic objects or behavior (fame, success, or prestige) like a car or money. There are several different non-moral values: 1) utility-good because of usefulness, 2) extrinsic-good because of a means to do good, 3) intrinsic-good in themselves, 4) inherently good to think about them, and 5) contributory good-contribute to the intrinsically good life. However, we did not learn the importance of these non-moral values. These non-moral values (money, fame, power, and success) are the means to the good life, the good extrinsic life. And, there is nothing wrong with wanting to be successful or to make money. Most of us want to be in this real world of things. However, there are some concerns. Again, I have to refer to Dr. Stoll's example to explain non-moral values. Your car has utility value—it is useful; it gets you from point A to point B. Your car has intrinsic value—it has meaning to you, because you always wanted a "car." Your car has extrinsic value—you can sell it and make some money on it. Your car has inherent value—maybe it was Grandma's car, and grandma means a lot to you, and whenever you see the car you think of the good times at grandma's house. Contributory value is all of the above. Non-moral values help motivate or move us to play, recreate, dance, or exercise. They are the reason that we participate and play from the social construct.

Non-moral values drive every moral decision you may make, and they can get out of hand. For example, if you choose to cheat in class, it is probably because you want a better grade (a non-moral value). If you chose to cheat on your boy/girlfriend, it is because there is someone or something you want, instead. If you chose to steal, it is because you want the money or object. Thus, the non-moral values begin to drive our decision process. How important is fame, power, success, and money? Are you willing to lie, cheat, and steal to get them? Are you willing to be disrespectful or irresponsible to get them? That is when moral reasoning comes into play.

MORAL VALUES

The word moral comes to us through the Latin word for character, custom, or manners. It refers to character or disposition, which can be good or bad, right or wrong, virtuous or vicious, or to the distinction between right and wrong in relation to actions, volitions, or characters of responsible people.

Character is how you treat others. Basically, morality is common decency to others. How do you treat others? Aristotle once said that character is how you treat others even when no one else is looking. Thus, Aristotle agreed that it was about motive, intention, and action. Unfortunately, most of us are mercenary, which means that we usually are doing good only for what we get out of it. Most of us have problems with good motives and good intentions. Therefore, the right actions are probably for mercenary reasons. However, we should be better. We should be better students. We should be better professionals. We should be better at being moral folks and treating others with respect and dignity. The question is: Are we willing to do the work to be better people?

Something's relative worth is determined by culture. This value is ascribed to persons, things, events, experiences, and so forth. The value can be moral (subjective, intrinsic) or non-moral (objective, material, extrinsic, instrumental).

Moral values are worth or importance placed on intrinsic behavior focused or directed toward other humans. Moral values take into consideration the motive, intention, and actions that affect or influence others. When we do a moral action, there is always a motive and intention driving it. My major professor, Dr. Sharon Stoll, had the best example to explain moral values. Consider this: Joe is on the football team. One of his teammates gets into a fight on the field. Joe believes it is his duty to help his teammate. So, he goes to help his teammate (a good motive—helping). However, his intention becomes questionable, because he intended to go out and get into a fight—bad intention. The resultant action, which is wrong, is that Joe, after putting on his helmet, jumps into the fray. Also, consider a slightly different example from Dr. Sharon Stoll to explain moral values: Joe might be a very complimentary guy. He tells all the ladies that they are beautiful—right action. Except, he only does it to get a shot at getting in bed with them—bad intention. And, sometimes it works—wrong action. In Joe's case, character is that condition in which people judge you. It is not your personality, you might be an exceptionally complimentary sort, like Joe, with charming personality, but you, like Joe, might be a womanizer.

PRACTICING GOOD ETHNICS IN MOVEMENT

Remember, we learned in Chapter One that philosophy is divided into four branches of philosophy: 1) metaphysics (the nature of reality—why do we play?), 2) epistemology (what do we know when we play a game?), 3) axiology (what do we value?), and 4) logic (the science of language). In this chapter, we will only focus on axiology, the study of value.

Axiology specifically asks questions of value. A value is something of worth. We place worth on an objective or experience. Worth is dependent on you. Does it have worth to you? And, how affected are you by what others think about worth? All of us are affected by culture and society in how much worth we place on something. Again, I have to quote Dr. Sharon Stoll, because she is an expert in moral reasoning. She says that, in our society, most young people own a car. When we describe our car, what adjectives do we use? Do we use words to describe color? Do we use words to describe its style? Maybe, if we include a certain car type. The bottom line is that most of us will describe our car by the make, model, and year (e.g., a 2006 Subaru Impreza).

My point here is that we are affected by our culture in how we describe the car. Rather than say it has four wheels, an engine, and a chassis, we tend to give its pedigree, because we want people to know its worth ... its societal worth. Pick-up, four-door, sport coupe, brand name, power, so forth and so on. It is not wrong or bad to describe your car with these adjectives; it is just the way we are educated by society to value cars. In fact, we are educated by society, friends, family, and media, which places worth on something, whatever that something might be: cars, clothes, money, and so forth.

Another example is our values about what is beautiful. What is beautiful? Would you say that lifting weights is beautiful? You might, if you have learned to value lifting weights. Would you say obstacle course racing is beautiful? My point here is that the values of beauty are learned and cultural. The aesthetic (pleasing to the senses) is learned, so what is beautiful or what is ugly is a learned perspective that is influenced by our experiences, by where we live, by who we hang out with, by our families, by our education, and so on and so forth. My friend, Andrew used to say that we are who we hang out with. But, since he is fluent in Spanish, he would also add: "Dime con quien andas, y te dire quien eres." So, know that what you think is beautiful might not be enjoyed by anyone else, or maybe, it will depend on where you are. To summarize, aesthetic values are personal, learned, and cultural. How we view values and what we do with them is very important, for they help us focus on what a culture (our culture is kinesiology—the study of movement) may think is good or what will bring us goodness.

VALUE IN SPORT

Non-moral values are usually objectively focused, which may be internalized also. These non-moral values are things that make the sporting experience better. In sport, non-moral values are usually objectively focused. The win. The applause. The media attention.

PROFESSIONAL ETHICS IN MOVEMENT

You are on your journey to become a professional in kinesiology, which is defined through our professional duties of teaching, leading, and serving in the disciplines of sport, physical education, recreation, dance, and health. We in kinesiology have always struggled with the differences between it being an academic discipline and a profession. It is the scientific and humanities knowledge that defines what we know. In contrast, a profession is about a practice.

WHAT IS A PROFESSIONAL?

A professional in the usual sense of the word might include all sorts of people in all sorts of jobs: a professional wrestler, a professional taxi driver. In this usual sense, a professional works full time for pay, has considerable expertise in what others may do for a hobby or a diversion, for example professional athletes. However, because someone is, in the generic sense, called a professional, does not mean that he/she belongs to a profession.

A profession and the professionals attached to the profession are different. Both profession and the professionals have access to and control over a particular body of knowledge. I am a professional teacher/professor. As a teacher/professor, there are certain responsibilities that I have. For example, I must make sure that the information that I teach you is current and research-based. The qualities of teaching—knowing how to teach and what to teach—is under my access and control but regulated by the university. A professional in a profession also performs a service to society. Teaching, coaching, leading, and dancing performs a service. As a teacher, I would say that the service I provide is inspiration to learn, which changes the course of the future. How do you see your service? A professional also has autonomy, which means I am in charge of my own destiny. I self-govern to a point. And finally, a professional in a profession is supposed to have prestige, respect, and social status. Yes, it is true—coaches seem to have a lot, teachers not so much, leaders, depending on how good they are, might have a lot prestige, respect, and social status. Dancers may or may not, which may be the difference between teaching and dancing … how much respect is given.

PROFESSIONAL STANDARDS

Originally, long ago, the available professions were thought to be: military, clergy, medicine, and law, because they had specific standards and codes of ethics and they were held to a high degree of professional standards. For example, the original professionals did not work for money, but for the good of people. Physicians, not ours today who work for corporations, were bound to the Hippocratic Oath. The oath basically tells us that the physician has special obligations to all fellow human beings. Professionals are not concerned with time clocks but work for the good of society.

PROFESSIONAL CODE OF ETHICS

Because a profession governs itself, there are usually codes of ethics written by a professional organization. I am a member of the National Association of Strength and Conditioning. The association has a code of ethics. Here, at Washington State University as a professor, I am operating under a code of conduct.

A code of ethics is written by the professionals within the profession, they monitor and decide who can become a member, and they set rules that members should follow. They also adjudicate if someone is accused of not following the rules and, then, can expunge that individual from the society.

A code of conduct, in comparison, is an institution that sets rules of behavior that an employee must follow. As a professor, I should not sleep with my students or commit other immoral acts with them … or lie and cheat, undermining the process of the institution, plus many other rules. If I violate these rules, I can be fired.

A code of ethics is not a code of morality, because a code of morality is a personal, private standard that one sets. A code of ethics for a profession comprises the acceptable standards of an organization. A code of ethics will not tell you how to live your life or how you should or should not live your life. However, it will tell you that you cannot have sex with your clients. A code of ethics won't tell you that it is wrong to sleep with your neighbor, but it will tell you it is wrong to sleep with your neighbor if that neighbor is one of your students, your clients, or your colleagues.

The purpose of a code of ethics is that it is regulative, it sets rules of behavior. The code also protects the public interest by making sure you know what you are supposed to be doing. The Code is not about you. The codes are very specific and very honest in their portrayal of what you are supposed to do. The code, if it is any good, must be policed and police-able, meaning that other professionals keep an eye on your behavior and are willing to expunge you if you do not do the right and proper thing. Most times, the right and proper thing has to do with competency, knowledge, and treatment of clients.

As I mentioned earlier, the purpose of codes of ethics, or codes of conduct, is not about making you a moral person, rather it is about making sure that you serve the public honorably and that you know how to and act as a professional should act in your field. A code does not guarantee that you will not make a mistake, but it does let the public know that you have been educated in a certain way and that there is a certain way that you should act.

WORKING WITH OTHERS

As an athlete or an active individual, we should have obligations in a movement community. What are some benefits we can gain from a movement community? What do we owe to a movement community? In fact, we are our movement activity. Movement is a practice, and we are among its practitioners. The appreciation and preservation of our movement activity depends on our fellow movers and us.

When I was in graduate school, I read a book named *The Philosophical Athlete* by Heather Reid. Reid (2002) asked some pretty intriguing questions about equality and difference in sports. For example, Reid (2002) asked if social and cultural ideas of inequality affect sport and how equality could be promoted within sport. The thing is that stopwatches and goal lines are inherently colorblind in the "judgments" they make our movement character. Sports have helped so many of us in our personal battles against social and cultural bias.

SHOWING RESPECT

Throughout my numerous encounters with sport and movement philosophy in graduate school, I have learned that sports do not build character, people do by thoughtful engagement in movement activities, such as sport and exercise. According to Reid (2002), to show respect, we first have to understand the Tripartite Theory from Plato's *Republic*. The Tripartite Theory consists of *logistikon*, the wisdom-loving part; *thymoedies*, the honor-loving part; and *epithymetikon*, the pleasure-loving part. I hope that most of us would not want to be like some gold medalists who use performance enhancing drugs and exploit their teammates, friends, and family. The truth is that great athletes overcome their shortcomings.

If you had to select players for your team, whom would you pick: a player who scores high in all categories but has the ability to work well with others? What is more important than ability to work well with a team?

THINKING ACTIVITY

The purpose of this thinking activity is to gain understanding in codes of conduct and codes of ethics in professional life (Stoll, Sharon, personal communication, August 13, 2012).

All's Fair . . .

You are the owner of a sophisticated line of women's clothing, especially in the production and marketing of a sport bra. Your line has been the best in the business for decades, and your company and you are known for quality products and business practices. However, in the past year, sales have declined due to what you perceive to be shoddy, yellow advertising by your closest competitor, BRZ lingerie. BRZ has maligned your reputation, as well as your established product, and now reigns as number one. As luck would have it, one of your designers brings you BRZ's fall production sketches. She states that, through a series of dumb luck incidents, she found your competitor's sketches in a designing class she was taking. After investigation, she realizes that one of BRZ's people apparently inadvertently left it behind. She excitedly notes that the material is dated, and it appears to be the latest model. She also states that, from what she can glean, your company can outdo BRZ easily and win back the lost market. What do you do, based on your stated principles?

- A.) Keep the model, tell your designer to be quiet about what she found and develop a new strategy based on what was found. Losers are weepers. All's fair in love, war, and the lingerie business. And obviously, BRZ has sloppy, as well as unethical, business practices, which now have caught up with them. It's payback time!
- B.) Tell your designer to return the model to BRZ, emphatically stating that you will have no part in clandestine nosiness.
- C.) Tell your designer to return the model, but only after you analyze it thoroughly. You're not a thief, but you're not stupid either.
- D.) Another option (please provide the reasoning behind it)

Let me share some of my thoughts on this philosophical/ethical dilemma:

EXAMPLE ONE

When confronted with the difficult situations in life, we find out who we truly are. Are we self-interested, even though we portray that we are not? Or, are our senses of right and wrong so central to our beings that self-demise is a better option than occasionally bending the rules? In reality, most people try to stay in that gray area between these two extremes. However, there will be times in all of our lives where a strong and passionate stance in one direction or the other is necessary. The hypothetical situation of being the owner of a sophisticated line of women's clothing versus BRZ lingerie presents a scenario where a staunch position is necessary. In my opinion, looking at BRZ's fall production model is an unethical and unwise choice that I would not make.

If I were to have looked at the model and developed a strategy from it, it would be devastating to my business. If anyone was ever to find out that I looked at BRZ's fall production in such a manner, they would be justified in thinking that I was immoral. I would not want this image, because it would not only be a slander against me, but it would also affect my already suffering business for a onetime offense. In addition, this would give credibility to BRZ's accusations of my supposed immoral reputation that they created. BRZ would also be justified in taking it one step further by treating my actions as an admission of my product's inferiority to theirs. This would surely be the demise of my reputation for quality business practices and quality products, a foundation that I believe helped my company reach success in the past.

Even if neither the public, nor anyone else, was to become aware of my sneak peak of BRZ's product, the action would have a negative effect on my perception of my business. Taking a sneak peak of my competitor's product would be extremely advantageous; however, it would be grossly unfair. If my company is unable to win a fair competition, than it is not worthy to be on the same level as my competitors. This remains true even if my competitors are using unethical methods themselves. My company is held to a higher standard, and I refuse to lower my standards to their level. This rigid thinking is due to my perception of what cheating is. Cheating is a strategy used by someone who cannot win a fair fight and is, therefore, a lesser to his or her competitor. Through this line of thought, the moment I took an unfair advantage over BRZ would be the moment I admit to myself that my company is a lesser to BRZ. Such an admission in my mind would be poisonous and self-destructive.

In keeping true to my morals while still being intelligent, I would destroy BRZ's fall production model without looking at it. I would not want to return the work back to BRZ for the likelihood of them assuming I looked at it. With BRZ's shoddy business practices, I would not be surprised if it tried to retaliate for my coming into possession of their work. I also propose that BRZ would feel justified for assuming the worst due to that being the likely course of action it would have taken. Rather, the results of my actions would be as if I had never come into contact with their fall production model.

EXAMPLE TWO

In law, a man can be guilty if he violates the rights of others. In ethics, he can be guilty if he merely thinks of doing so. In terms of professionalism, ethics and the law often coincide, but they also may deviate depending on what type of values companies and professionals maintain in a situation, like the hypothetical one presented by the prompt.

For example, the designer had innocently found the model from a competing company, but the consequent actions of the company depend on the values we hold. The four options (A–D) will be discussed in this reflection in terms of ethics and professionalism in business.

Choice A suggests that the company utilize the model and develop a new strategy based on BRZ's model for the sports bra. Rationale for this option includes the opinion that BRZ has had sloppy, unethical, and questionable business practices in the past, and finally, karma is catching up to that competitor. Likewise, our company suspects BRZ of shady advertisements that make our own company look bad. This option is placing priority on the values of justice and responsibility. For instance, BRZ deserves a bad turn for all of the negative practices it has maintained. Likewise, it was careless and left the model laying around, so it needs to accept the consequences for its lack of responsibility and thoughtless security measures. Also, if this model is beneficial, our company is looking out for itself to support the value of an aesthetic experience for its customers. The social value of determination to provide the best products is also a value that this choice might support. Non-moral values achieved by taking the model include the potential or regained fame, money, success, and prestige. However, even though this choice may support several well-meaning values, it is not honest and respectable professionalism in terms of how a company should act. The model was not designed by our designers and even if it is changed, credit needs to be given to the BRZ Company.

Choice B is the opposite of Choice A. For instance, the designer is told to return the model to BRZ and to cease clandestine nosiness. This choice reflects professional etiquette and the values of honesty, beneficence, and respect. Giving the model back without using it is a polite and honest action, even though using it might benefit our own company. Using the model also entails the possibility of a lawsuit if the designers at BRZ can prove that it is their model. Choice B shows that a good reputation, honor, and esteem are long-term goals that will benefit the company, and BRZ will ultimately respect our company more for being courteous even when we had the chance to use the model to benefit ourselves. Choice B also reflects the social values of hard work, because we refuse to use someone else's hard work and, instead, rely on our own dedication and abilities. However, this option might mean that BRZ will get ahead of our company so much so that we have to downsize or make major adjustments, and employees might regret being, perhaps, one of the only companies that is behind, because it refuses to resort to dishonest measures that other companies are willing to utilize. In this light, upholding these values in our professional organization could imply a negative result for our customers and employees.

Choice C offers a compromise between choices A and B. It upholds most values by giving the model back, but it still makes certain to include the possibility to analyze it just in case our company needs an idea or if we want to prepare for how the model might be used in the market. In doing so, this choice reflects most values, although it is still treading on values that align with total honesty. However, this model incorporates truth by showing BRZ that yes, we returned the model, but we know what BRZ is going to market and we know what the model is. By rights, BRZ was careless and our company could have taken the model and used it (assuming there was no formal patent for the design yet). This incorporates truth by reflecting that the model was not ours, but acknowledges that at least one of our employees knows the details of the model and might make use of those details later. This way, BRZ has to accept that our company is honest in respect to what could happen in the future, but was also truthful, polite, and professional enough to return the model that BRZ had left behind.

Other options that our company could consider (Choice D) include: Forming collaboration with BRZ to make the model even better and hiring the designer who actually developed the model. These options would incorporate all the moral values and still help the company succeed, because they are honest policies that offer more than just competitiveness or honest professionalism. Hiring the designer would make sure that we are using the work of an individual in our company, should he/she decide to share her non-BRZ–patented ideas with us, and collaborating with BRZ helps promote the company while still giving them credit for the model.

Getting caught for a crime under the law and negating ethics can be distinct from each other, and in the same way, professional ethics can vary depending on the values one holds. On one extreme, the company and its values for success, money, and security of its employees is most important, regardless of whether the choice is honest. On the other side, the values of respect, honesty, hard work, and good reputation are more important than competition to ensure that we are the best company. Clearly, this reflection showed how professionalism and ethics can conflict and coincide when choosing which values are most important, as it evaluated the aspects of several options when faced with a dilemma in a professional situation.

CHAPTER SEVEN

The External and Innate Importance of Movement

As movement professionals, how can we help others build physically active lives? Health, fitness, knowledge about movement, motor skills, fun, and pleasure may be able to promote other beneficial things in our lives. If you could only have one of these values to help others achieve a better life, which one would you retain and why? How can we enjoy our movement lives, if we are unable to experience fun and pleasure? In this chapter, you will learn about the extrinsic and intrinsic importance of movement related fitness/health, knowledge, movement skill, and fun/pleasure. This chapter requires us to ask a series of questions:

- Do we have to experience and appreciate movement in order to live a satisfactory life?
- What can we suggest to those individuals who lead sedentary lives, but still claim that their lives are completely pleasing?

THE EXTRINSIC IMPORTANCE OF FITNESS/HEALTH

The extrinsic value of fitness in movement-related activities is that the more fit we are, the better we feel about ourselves. In addition, we tend to experience a self-confidence boost, which can influence our success at work, at school, and in our social lives. We can also experience optimistic self-identity, where we feel youthful with thin bodies and tight skin. After all, if we are not healthy, we may never achieve our goals in our life.

Often times, we will experience discomfort, or maybe even some pain, when we engage in physical activity. Quite a few people are not comfortable with experiencing pain when they move. Also, fitness and health can be financially demanding and boring for some individuals. The point of fitness and health is that it requires thorough maintenance. We can work for months to improve aerobic endurance and lose everything in a matter of week or two.

FITNESS/HEALTH VALUES (KRETCHMAR, 2005)

Aerobic Endurance

Aerobic endurance is the ability to move dynamically using a somewhat hard, challenging intensity that triggers energy expenditure. Aerobic endurance allows us to work on any daily activities, like cleaning the house, running to catch a bus, or trying to make it to our classes on time.

Anaerobic Endurance

Anaerobic endurance is that sudden and explosive burst of energy that helps us to complete our daily tasks quickly and efficiently. Due to a highly explosive nature of anaerobic endurance activities like sprints and weight lifting, We do not relay on oxygen to complete any anaerobic tasks. The best example I can provide for anaerobic endurance is that this type of fitness is needed to help save people in a physically taxing emergency (p. 234).

Agility

Agility involves the ability to change directions in any sport, reacting to avoid opponents, and executing plays gracefully and quickly. This type of skill can help one achieve success and fame in sport (p. 234).

Body Composition

Body composition is the ratio of your fat mass and fat-free mass. Normal body composition (approximately 20%–24% body fat for females and 14%–18% body fat percent for males) allows you to move effortlessly and increase your chances of a long and productive life. Body composition may as well be one of the strongest extrinsic motivators, because of its ability to improve physical appearance and increase self-esteem.

Muscular Strength

Muscular strength is the ability to lift the maximum amount of weight by an individual in just one repetition. Muscular strength is extremely significant in daily living, because it offers us the ability to help us lift things. It can promote independent living, especially in older adults (p. 234).

Flexibility

Flexibility is the ability to move our joints through a full range of motion. Good flexibility helps with chores that involve bending, reaching, and stretching. Maintaining/improving flexibility can improve independence in older adults (p. 234).

THE EXTRINSIC IMPORTANCE OF MOVEMENT-RELATED KNOWLEDGE

The more we know about different ways to positively influence our longevity, the more comfortable we feel about being able to lead a longer life. We can also gain a deep understanding of our bodies to prevent injuries and increase recovery time. With more knowledge comes more understanding on how to increase sports and/or exercise performance, promote sport and/or exercise participation and retention, and experience more victories in games. It is important to mention that knowledge can also help with competition, evaluating progress and play strategy, understanding techniques that can lead to success and prestige, and understating rules, current issues, and fans.

THE EXTRINSIC IMPORTANCE OF MOVEMENT SKILL

There are several movement skills that will be discussed in this chapter. Sport skills are extrinsically important, because they can bring fame and financial stability. Dance skills can help individuals express themselves through movement. Exercise skills can help with health improvement and overall physical fitness. Recreational skills can increase fun and help individuals avoid boredom. Overall, individuals can truly express themselves through body language, explore new movement creations, and experience new movement methods.

THE EXTRINSIC IMPORTANCE OF FUN/PLEASURE

Many of my former and current kinesiology students would argue that one must have fun in order to adhere to exercise and physical activity. If the activity is not fun and engaging, people will not continue. And, if we do not continue to be physically active, we may never experience the fun and pleasure of movement. Kretchmar (2005) explains that individuals can experience different types of fun/pleasure (p. 236–237).

1. **Test fun** is the ability to solve problems and get satisfaction from those solutions. The benefit would be to play a challenging game, but still get some pleasure from it. I teach group fitness classes, and I have come to realize that majority of my participants like my Zumba® class the best. Zumba is a dance-based aerobic class with different Latin-based international rhythms, which can be very fun and engaging. I do not have any experience with dancing, so my Zumba classes tend to be more fitness- and sport-oriented. Often times, I do like to attend a Zumba class from a particular instructor who used to be a dancer in college, because it is difficult and challenging for me to get all the movements, but still very enjoyable and fun. It is challenging to know how I will do when I am taking Zumba classes from dance background instructors.
2. **Contest fun** is the ability to perform better than everyone else. The benefit of this fun is to experience the closeness of a hard-fought game, appreciating victories and opponents, and finding satisfaction in future challenges. This appreciation and satisfaction promotes performance, success, and fame. This alone promotes performance, success, and fame (p. 236).
3. **Sense fun** (sedate) comes from being comfortable in our body/self experiences and enjoying how our bodies feel after a movement activity of choice. Enjoyment part of fun can bring us back to those feelings over and over again. This type of fun can bring us back to those feelings over and over again.
4. **Sense fun** (dramatic) is achieved when we experience the thrill of skiing down a mountain, attacking our opponents while playing good and fair defense, and going fast on our bicycle. A thrill while performing a certain activity is a huge component of sense fun.
5. **Aesthetic fun** is that enjoyment that comes from beauty of movement. It can promote a sense of tranquility and existence. This beauty can bring more meaning to a certain activity which in return promote a sense of tranquility and existence.
6. **Ludic fun** is all the experiences that we appreciate for their own sake. For example, we may appreciate trail runs, not because we are forced to, but because we enjoy trail running. These types of activities lower stress and develop consistency in our life.

The motivational power of fun/pleasure is easily understood. Most of use would agree that attending parties with enjoyable people is fun. As for physical activities, they all have to be fun and pleasurable in order for us to be involved with movement and return to movement. In fact, we often abandon everything that is not fun sooner or later. We also have to be careful about what type of movement we prescribe to our participants. If movements in any physical activity of choice become too challenging, our participants may experience anxiety and never return to their workouts. Any movement professional that removes fun from physical activities is looking for a disaster to happen. Kretchmar (2005) summarizes this best by saying that we have to build something that is meaningful into our participants and something that connects our past to our present and our present to our future (p. 238).

If we can attempt to prioritize fitness/health, knowledge, movement skill, and fun/pleasure and give at least one reason for our ranking of each, what would the order be? Which one is more important and least important, and why?

INTRINSIC IMPORTANCE OF MOVEMENT

If I were going to ask you a question about what motivates you intrinsically to be physically active, what would your response be? The problem is that intrinsic motivation is really hard for people to understand, especially sedentary and people that are no physically active. When we first introduce our clients/patients on a exercise program, we need to be able to give them something that they can extrinsically motivate themselves with before we introduce the intrinsic importance of movement. Once they establish that intimate relationship with movement and become hooked, we may be able to try some intrinsic motivators. Satisfaction is experienced not because we appreciate our physical appearance, but because we cannot imagine our lives without movement. People develop movement stories and meaningful narratives that they can share with others in the hope that they will at least try to experience the benefits of movement and physical activity.

THE INTRINSIC IMPORTANCE OF FITNESS/HEALTH

Have you ever experienced being alive, powerful, and full of positive and contagious energy while exercising, playing your favorite games, recreating, or hitting the skiing slopes? You feel light and worry-free and in the moment where all the movements flow effortlessly. Kretchmar (2005) would argue that we do not live for health, but live from it (p. 235). We do not develop our personal narratives around fitness/health, but we find meaning on the basis of whatever level of fitness we currently have. Ask yourself this question: "Does it feel better to experience an extraordinary fitness level, be adventurous in our travels to new places, or be in love?'

THE INTRINSIC IMPORTANCE OF KNOWLEDGE

We are extremely concerned with how our bodies work and how they feel. We devote our time and energy into learning new skills and usually compete against ourselves. When we do choose to compete against others, we are often just testing our knowledge. Ask yourself if knowledge is the prime value of movement. Do we need to have a basic knowledge about certain movements in order to engage in those movements? What are we striving to learn from our movement activity?

To produce a healthy life, knowledge must be satisfying and meaningful. Often, we have to stop and think if we would rather know the factors that lead to excellent performance or embody them? Do we get more satisfaction from knowing that we are better than our opponents or we need some component of doubt and tension?

> There is no value we can sell as well as knowledge. People spend millions of dollars on how-to books and literature ranging from topics like diet and exercise to stress reduction through sport. People want to know how to live longer, how to say young, how to win Olympic gold medals, and how to dance and play better.
>
> (Kretchmar, 2005, p. 213)

THE INTRINSIC IMPORTANCE OF MOVEMENT SKILL

When we master our movement skills of interest and choice, we feel at ease and in harmony. We experience creativity and peace, and we are truthful about our movements. There is a sense of freedom that comes with skill development in movement. Kretchmar (2005) discusses five different human freedoms related to intrinsic value of these skills (p. 242–244).

1. Freedom to discover. As we master certain movement skills, we are freer to explore new movement discoveries. As our cardiorespiratory fitness improves, we are, for example, freer to experience longer-distance trailing running adventures.
2. Freedom to explore. If we have stronger explorative skills, our discoveries will be more meaningful and richer.
3. Freedom to express. Eleanor Metheny (1972) once said that movement is "a flow of meanings without speech" (p. 226). Through this freedom to express, we find a way to tell our movement stories to the world.
4. Freedom to invent. Movement skills can help us invent new things in physical activity that are creative and challenging. People who are inventive in solving problems experience a freedom that is not available to others (p. 243). No one thought it was possible to beat a 4-minute mile until Roger Bannister did it.
5. Freedom to create. Art can inspire us, but when the magic is broken, we are left with sounds, colors, shapes, and patterns. When we are stirred either as creative performers or as spectators who appreciate art, we are appreciative (p. 244).

Individuals who are afraid to express themselves through movement or who do not move well are often afraid to participate in any physical activities. Try to watch your movement educators/instructors/personal trainers/athletic trainers teach movement. They teach movement culture. They prepare kinesiology students to enter a society, complete with its values, stories, facts, and myths (Kretchmar, 2005, p. 214). It has been said that if you want to understand individuals and discover who they really are, you should watch them when they move, play games, or participate in any physical activity. If we can create a meaningful and personally significant movement activity for our participants, we have won!

THE INTRINSIC IMPORTANCE OF FUN/PLEASURE

Pleasure, in movement philosophy, is the experience of aesthetic joy and satisfaction from any accomplishment in our movement activities, satisfaction because of good play, and the experience of meaning behind our movements. If I a master a lifting technique I have been trying so hard to master, I will achieve some standard of excellence and a lot of pleasure. The value of this excellence is uncertain until I know its significance for me and what it means in my movement life. If I gain this excellence as part of a lifelong passionate goal, that is one thing. But if I gain it as a result of a pushy coach and parents, that is quite another thing. Here is an interesting question we can ask others and ourselves: "Why is that so many children report unsatisfactory experiences in their physical education programs" (Kretchmar, 2005, p. 216)?

THINKING ACTIVITY ONE

Kretchmar (1993) provides an interactive philosophic exercise that I often use when I teach my movement philosophy classes to get my kinesiology students to think in a different way. He combines two distinctive pairs and asks which one you would choose and why?

Pair 1:

- Living 90 years
- Being an excellent and respected coach, teacher, or other movement professional

Pair 2:

- Living with a normal blood pressure and a general absence of pain
- Being accepted and loved by family

Pair 3:

- Having important goals for life
- Having physical and financial security

Pair 4:

- Having an enjoyable life
- Having a fit life

THINKING ACTIVITY TWO

To be a movement-educated professional you must embody several characteristics, according to Kretchmar (2005). A movement-educated person embodies:

- Competency in motor skills and movement patterns needed to perform a variety of movement activities;
- Understanding of movement concepts, principles, strategies, and tactics as they apply to learning and performance of physical activities;
- Participation in regular movement activity;
- A health-enhancing level of physical fitness;
- Responsible and respectful personal and social behavior in physical activity settings; and
- The benefits of physical activity such as health, enjoyment, stimulation, self-expression, and social interaction (p. 209).

Do you embody these characteristics? Can you come up with more characteristics a movement-educated professional should maintain?

THINKING ACTIVITY THREE

Why did you choose to study kinesiology? Kretchmar (2005) argues that individuals who fail to prioritize extrinsic and intrinsic values are often seen as lacking as movement professionals. Here are some of values you should embody as a movement educator/professional.

- Know what you stand for (practice what you preach as a movement educator/professional)
- Know and understand your commitments toward physical activity and movement

- You are not just studying kinesiology to make a living (you are studying because you understand the importance of sharing your physical activity and movement experiences)
- You know which direction you will take (in explaining your physical activity and movement experiences)
- You have courage to express your opinion (be confident in your knowledge about physical activity and movement)

CHAPTER EIGHT

The Importance of Play and Games

This chapter will examine the importance of games (including sport) and the importance of play. Before we start discussing games and play, ask yourself why games exist in every culture around the world. In his first edition of *The Practical Philosophy of Sport*, Dr. Kretchmar (1993) posed an interesting question about how something can be "only a game" and yet generate million-dollar contracts, cause large stadiums to be built, and attract hundreds of thousands of fans?

THE IMPORTANCE OF GAMES

Playing a game is the voluntary attempt to overcome unnecessary obstacles (Suits, 1972, p. 22). Kretchmar (2005, p. 162) explains that, to understand the importance of games, we need to ask ourselves the following questions: What is the game about? What is the goal? What are we allowed and not allowed to do in pursuing the goal? How does the game start and stop? What counts as a complete game? What happens when a rule is violated, equipment breaks, unusual weather conditions intervene, or other forces interrupt the game? How do we get a compromised game back on track? How would two or more people share this game?

In a game, at least two participants take the same assessment, so that their scores can be compared in a meaningful way. In addition, all participants must promise that they will perform their best. These assessments do not need to be competitions. Assessing one's technique and form while lifting weights in the gym can be significant, meaningful, and interesting and valid assignment.

Sport philosophers would argue that games are more intellectual than play, and according to Kretchmar (2005), this statement requires two subjective advances it requires two subjective advances:

1. An interest in embellishing a challenge—that is, the ability to appreciate that an unnecessary problem may actually be a good thing (p. 164).
2. The intellectual ability to connect the means and ends of this artificial test—the capacity to see, for instance, that achieving the end by alternate means does not count (p. 164).

The first advance is about the ability to appreciate logical problem solving. The second one is about appreciating the meanings of rules and relationships with other participants. Simply put, without any challenges our problem-solving abilities diminish, and without movement or skills suffer.

There are two different types of games: shallow and deep (Kretchmar, 2005). Shallow games grab our attention for a short time. The problem-solving ability is desirable, and we take on shallow games with excitement, but, if we cannot solve them, we get frustrated. If we play these games well, we experience immediate satisfaction, but we are not likely to go back to playing these games again. On the contrary, deep games can be very addictive. The problem-solving aspect of deep games is very complex. As Kretchmar (2005) explains, deep games require good fortune, skill, and insight (p. 168). These deep games are very significant and meaningful. They are concerned with relationships, with oneself and with others. As a power lifter who reaches a personal goal of bench pressing 345 pounds is not done. The power lifting problem is still there. What about bench pressing a 400 pounds or 450 pounds? Improved bench pressing technique does not pose any threat. What if we get addicted to anabolic steroids to achieve those goals of bench pressing 450 pounds?

Games should promote success, opportunity, and justice (Kretchmar, 1993). Success is good (one is by winning/reaching a personal goal/showing improvement). Freedom (opportunity) is good (everyone has at least a narrow opportunity to succeed, to improve, even if the game cannot be won, the average rookie stands at home plate under the same testing conditions as the future hall-of-famer). Justice is good (everyone is expected to play fair). We find out what we like, what we can do, how we react to pressure and adversity, how we relate to friends and opponents, what our capabilities are, what gets our attention, and where we like to be.

THE IMPORTANCE OF PLAY

The issues of seriousness and importance help distinguish games from play. True players live for playing, are attracted by the possibility of participating, and give little thought to outcomes or consequences that do not affect the quality of the experience itself. When you truly play, you miss dinner, court physical disaster, encounter exhaustion, and get caught doing silly things. Players find it difficult to stop their activity and return to the real world. Play is freely chosen (Huizinga, 1950). The more we aim at orchestrating fun while playing, the less likely we are to experience it, at least in its full form. Play allows meaning to count (People say, "I could have danced all night" or "I wanted to play golf all day"). We rarely hear people say, "I wish I could have exercised all day" or "I could have displayed healthful posture all night." Often times, we associate play with children. One interesting thing about children is that children will not do activity simply because it is good for them, but they are active, because they enjoy it (Corbin, 2002).

Like shallow and deep games, play can also be categorized as shallow and deep. Shallow play mostly comes from intrinsic factors; it does not inspire creativity. Deep play is very individual; it becomes a part of us. For example, when you truly appreciate a type of play, you skip meetings so you can play that activity, you read and educate yourself about that activity, and you watch and learn from other participants. A runner who experiences a deep connection with running as a meaningful movement is no longer a slave to running, but running is who he/she really is. If this deep activity becomes too deep, it can be unhealthy.

> Values change, worldviews change, skills change, spaces and distances change. Hard becomes easy, far becomes close, the mechanical becomes expressive, someone else's challenge becomes your challenge. The playground has become part of the player.
>
> (Kretchmar, 2005, p. 156)

THINKING ACTIVITY

In groups, discuss the following questions:

- Do you still play regularly? If so, where and how?
- Can too much play be harmful? If yes, how?
- Can kinesiology professors/instructors/teachers promote a spirit of play and still help students achieve other good items, like getting physically fit, staying healthy, and developing good character? If yes, explain how.
- How do you play games: as if they do not much matter as if games were all that counted in life?

- In this performance-oriented culture do you think that some professional athletes have grown tired of their games and now play for the money? Name some athletes that do this.
- Do you think that some insecure children and adults cannot enjoy sport itself, because they are too worried about failing and losing? Explain.
- Do you think the kinesiology field promotes fun and games?
- Explain how we can promote movement? As work? As play?
- Would it be smart to ignore games and emphasize only movement and/or exercise?

CHAPTER NINE

Competition in Movement (Sport, Exercise, and Physical Activity)

In this chapter, we will address competition in movement, discuss winning, craft a definition of an athlete, and understand the importance of competition in sport, exercise, and physical activity. Through my graduate school classes in sport and exercise sciences, I heard many definitions of an athlete. The one definition I will always remember is that the word "athlete" in Greek means "to contend for a prize," which would mean that we have to have some standard of excellence involved. To explain competition even further, we need to elaborate on the above definition of an athlete. Competition, in fact, is man's struggle within life for personal excellence. An opportunity for victory through the struggle provided by the contest or battle is one's ideal of perfection. I could not help but think about developing my own definition of an athlete when I was writing my doctoral dissertation. Here is my definition of an athlete, what is yours?

> An athlete may be that person who persistently seeks new adventures, visualizes new movements, hears his or her mind and body, listens to his or her mind and body, anticipates, imagines astonishing body movements; an athlete who may be struck by his or her faith in their body and mind chain as he/she came from the me world where extraordinary movements just seem to flow, but may never suspects his or her mind, body, movement, play, exercise, and performance.
>
> (Mijacevic, 2012)

THE IMPORTANCE OF COMPETITION

In the previous chapter on games and play, we learned that movement need not be competitive. Our movement experience can, in fact, be very meaningful and significant without any competition involved. Why would we want to be involved in a physical activity that always promises a winner and loser? Kretchmar (2005) calls this philosophical dilemma moving from a non–zero-sum to a zero-sum activity (p. 170). He further explains: If you are getting a good score in a contest, it does not prevent me from getting a good score as well. If you are improving in a contest, it does not prevent me from improving too. Therefore, there is no zero-sum relationship between the scores. When this contest becomes competitive, your victory means my loss. You get everything; I get nothing (zero-sum relationship). Thus, why would we want to trade a win-win testing arrangement for a win-lose competitive project?

To further understand this zero-sum philosophy, think about when you were competing against your opponent and lost. Most likely, you were playing well and making the moves at just the right time, employing the right techniques and strategies, handling stressful situations, and still lost. In our competitive culture, we are taught that second place is a first loser.

Often times, we like to make our games, and even play, more fun by adding some competition. When we do that, does competition spice up a potentially boring activity and make it fun, or does competition create tension in the doing and make you look forward to the end result?

Competition in any physical activity "is like a little salt, it adds zest to the game and to life itself. But when the seasoning is mistaken for the substance, only sickness can follow" (Coakley, 1978, p. 62). Numerous studies on competition and athletes have shown that their sport participation declined, because they felt burned out from their sport activity. This finding led the researchers to conclude that these athletes did not learn how to enjoy themselves in organized programs; their focus was always on the outcome, rather that the process. What would happen if we stopped judging athletes on the basis of wins, losses, trophies, prestige, fame, and money and start judging them on the basis of effort, personal progress, moral values, and ethical decisions?

I have to be a little more philosophical about competition and talk about the aesthetic beauty of competition. What can be beautiful about competition? Opponents struggling against each other? Athletes struggling against themselves? How well you do in competition is how well you struggle. If you do not struggle well, you should feel bad (not when you lose). When we play against other people, we often merely seek out the experience of a challenge (not a win), testing our skills, and adding another dimension to our involvement.

WINNING

We are taught to believe that winning or success is the objective for every game. Our desire to win is often very central to our movement experience whether is in sports, exercise, games, and even play. Usually, we are very hard on ourselves, and we do not realize our full potential by losing. We have to ask ourselves: Whom do we conquer by competing? None, but ourselves? Have we achieved the ultimate satisfaction by winning?

In sports, there are more losers than winners. There must be more to sports than establishing superiority. Vince Lombardi once said that winning is not everything; it is the only thing (Stoll, Sharon,personal communication, February 12, 2009). But, Vince Lombardi also described a champion as that athlete who plays well whether they are losing or winning. Winning should be our ability to give total energy to be the absolute best we can be. Professional athletes, for example, are paid and measured by their final score but under looked on their total effort against their opponents.

Competition is a studied phenomenon, and "people are not born with a motivation to win or to be competitive" (Eitzen, 1984, p. 136). Humans adapt an activity with an instinct to survive where the primary reason for contest is the outcome. Sport may as well be a reflection of survival of the fittest. This winter, I hiked the Moscow Mountain and it took me 5 hours, but I made it all the way to the top. The thrill upon arrival on a mountain peak would be without much meaning if I had decided to ride a four-wheeler. In this sense, the Moscow Mountain summit is the symbol, and attaining it is confirmation of obstacles overcome. Both competitive athletes and recreational athletes share a common wish to confront physical obstacles, not just using their physical abilities, but the qualities of their character, as well. How would you define competition? In terms of winning and losing? In terms of other sources of enjoyment? Is the occasion for competition more important that the outcome?

PLAYING FAIR

Fair play involves taking a stand beyond the rules of the game. Would you rather take a stand that places your winning at risk or a stand that preserves the dignity and value of sport? Think about these statements: To win by cheating, by an umpire error or by an unfair stroke of fate, is not really a win at all. Or, a win is a win, no matter how it was obtained.

If I were going speak from my own personal experiences, I would say that playing fair is important to me, but winning is also important to me, but what bring me more joy is the experience of being fully engaged in whatever movement I am performing. I get unhappy when my mind begins to wander during movements. If I do not feel completely connected, sometimes the defeat will make me feel better than a victory. When competition is severe and strenuous, playing fair is often forgotten, rules are violated, we attempt to injure our opponents, we perform some underhanded practices, and we tend to forget to remain focused on the athletic event, but focus only on our potential victory.

THINKING ACTIVITY

If we were rational machines:

- Everyone would use seat belts
- Everyone would have stopped smoking or not begun in the first place
- Nobody would take undue risks
- Trainers could get everyone exercising simply by explaining all of the health benefits

It is far more important to hook people on activity! What are your ideas?

CHAPTER TEN

Self-Knowledge and Discovery in Physical Activity

One argument that I would like to mention is that movement was available to us from the start. Before we even learned any movement activity that movement activity was already there. For example, kicking a ball, lifting weights, and catching were already there before we learned them. Indeed, movement is at the core of every individual's engagement with the world, because it is in and through movement that our existence acquires reality (Husserl, 1989). In effect, movement forms the I that moves before the I that moves forms movement (Sheets-Johnstone, 1999).

In terms of experience and most importantly, we begin our life with movement (stretching, wiggling, reaching, and opening our mouths). It is interesting that most of us (scientists and philosophers) overlook this phenomenon and fail to explore its importance. We literally discover ourselves in movement (Sheets-Johnstone, 1999). We discover that our knees can flex, our core can tighten, and our shoulder joints have mobility. Our movement is freely variable, qualitative, and a rich measure of kinesthetic consciousness.

> What is already there is movement, movement in and though which the perceptible world and acting subject come to be constituted, which is to say movement in and through which we make sense of both the world and ourselves.
>
> (Sheets-Johnstone, 1999, p. 138)

Consider this argument: "Kinesthetic motions are the most fundamental dimension of transcendental subjectivity, so that even the body as functioning body, is not just something constituted but is itself constituting as the transcendental condition of the possibility of each higher level of consciousness and of its reflexive character" (Landgrebe, 1977, p. 108). This statement argues that our movement experiences are the most fundamental aspect of functioning bodies. To better understand this argument, Husserl (1989) explains that the body is a freely moving organ through which the subject experiences the external world.

In fact, whatever type of movement we do, whether we lift, pull, push, contract, relax—we do so in one piece. Our whole body is engaged in movement, sometimes engaged simply by being still, as in the preparation to lift an object. This phenomenon where body moves as an integrated whole can very well be harmonious, and we can be completely conscious and aware of our movements.

Now, you will experience that, often times, you cannot put your movement experiences into words. When somebody asks you to describe how you were so fluid in your lifting technique or your long jump, it may be difficult to find adequate adjectives or nouns by which to describe the different qualities we experience in moving. While the words may be hard to find, the qualitative experience itself is kinesthetically distinctive. We can conclude that we make sense of our bodies through movement more often than we do so with words. Through movement, our bodies are the very source of our being in the world and the center of our kinesthetic experiences. Evidently, when we focus our attention away from everything else around us and toward the movement of our bodies, we experience ourselves kinesthetically, and we recognize our movement.

In the beginning, we all had to learn our bodies. None of us is thoroughly knowledgeable of everything our bodies can or cannot do. Through our movement experiences, we learn how to discover our bodies and not how to control and manipulate our bodies. For example, picture an infant, and ask yourselves if he/she trying to control his/her body or if his/her body is an out of control body waiting for a mind to catch up with it? As an adult, what is it like to learn one's body by simply experiencing it, rather than attaining the mind over matter philosophy? How do we learn our bodies from our movement experiences? The simplest answer is by moving, and through this movement we challenge ourselves to learn our bodies.

How can we experience the most purely kinetic and natural movements, like walking and stretching, and move to more complex kinetic experiences, like proper lifting techniques, running techniques, and dance? Purely kinetic movements have no actual goal or purpose. In fact, we are not just walking to get someplace; we are not stretching to increase the range of motion, we do it because movement itself is significant and meaningful.The true meaning of our kinetic experiences is in the movement itself. Through these experiences, we may be able to discover our kinetic/moving consciousness and what is it to be a mover. Whether movement happens to us or whether we make it happen, when we focus on the experience and self-knowledge of a certain activity, we find the preciseness of our dynamic and kinesthetic experience.

What constitutes knowledge, in this case self-movement knowledge? Can we say that only the person who has mastered kinesthetic experience is able to obtain this knowledge? One must learn one's body and move before he/she can obtain this self-movement knowledge. Clearly, if one were really to know everything there is to know about the physical nature of the world, then one would, first of all, have to experience oneself as a moving, kinesthetically sentient creature (Jackson, 1991, p. 392).

With each movement we make, we should be developing personal stories and unique experiences. We must truly understand that movement is a phenomenon, that "I move" (fully with both mind and body) is essential to movement. The question is, "What do we know about moving body/selves?" What we do know may not be what we think we know. In this case, we need to reflect on our moving experiences both physically and mentally. First of all, a brain is not a body. To collapse physical into brains is an epistemological mistake (Sheets-Johnstone, 1999, p. 213). Thus, moving bodies contain brains just as they contain hips, knees, and ankles. A brain is a part of body. Simply put, the ability to move is to be free and alive with the possibility of doing anything.

The ability to be physically active is our kinetic aliveness and liveliness. We are born to be kinetically engaged with the world around us. Whatever our initial motivations are, they are developed by our bodies' kinesthetic actions. Motivation is experienced and practiced. It comes from bodies' meaningful feelings and sensations. The argument is that the body itself it the object of inspirations and intentions—in the form of jumping, running, stretching. When we are engaged in any physical activity, we learn ourselves and learn our bodies. The learning process does happen through our kinesthetic experiences.

To further understand movement, let us examine the natural movement attitude in babies. Why are babies so drawn to movement? They learn their environment through their own tactile and kinesthetic bodies. Furthermore, they find meanings and values in their environment, which they experience through tactile encounters. Babies often perceive something other than words. They perceive thinking, not in words but in movement.

Our subjective physical activities are kinesthetic experiences of our own bodies and movements, and they are important to our self-discovery, thinking process, and knowledge. The fact that electrical stimulation of a brain is not sufficient to induce meaningful, coordinated movement is compelling evidence as to why a mind is not a brain and a brain is not a body (Sheets-Johnstone, 1999, p. 441). What does it mean to think in movement? As soon as man uses movement to establish a relationship with his fellow men, movement is no longer an instrument, no longer a means: it is manifestation, a revelation of an intimate being and of the psychic link that unites us to the world and our fellow men (Merleau-Ponty, 1962, p. 196). Caught up in our busy world, we easily lose sight of movement and our capacity to be physically active. Any time we decide to dedicate our attention to movement, movement happens.

CHAPTER ELEVEN

Refining Life Through Movement Professions and Writing, a Personal Physical Activity Philosophy

When we hold on to wonder to see if there is something intriguing and new, we may arrive at a philosophic turn. If we pursue and generate our thoughts, ideas, and/or opinions, we are involved in a task called philosophy. When we start to wonder about love, death, cheating, minds, and bodies, we strive to understand something about our lives. The relationship between a philosopher and his audience is serious and purposeful, not social, casual, or random. (Blumhagen, 1979, p. 111).

Throughout my experiences as a movement professional, I have often heard and read that movement, physical activity, and sports professions focus only on the physical and mechanical and that has nothing to do with ideas, knowledge, and understanding. Kretchmar (1993) explained in his *Practical Philosophy in Sport* that education must be intellectual, and we cannot make any claims that our profession satisfies this criterion. For example, the same individuals would claim that learning how to play games, or how to workout and be physically healthy is exactly the same as learning how to clean and fix things. In fact, these skills are very useful, but are they educational and intellectual? To continue with this explanation, these skills can help us keep our jobs, but they do not inform and educate. Therefore, we cannot satisfy the criteria that a movement profession is an intellectual profession.

Think about this statement: "What do we promote in our movement professions?" Do we tend to promote the development of repetitive habits? Our students are predictable like robots. When we do any movement drills, agility exercises with changing directions, or multi-joint movements, are they intellectual changes or are they just insignificant? Mind and body working together to execute any type of movement can be impressive and intelligent habitual. In this sense, are movement skills different from writing skills?

Think about your favorite athletes; are they just mechanical and predictable? Or, do they still foster a repetitive behavior by perfecting their movement habits and reflexes?

Movement and physical activity can be a dialog where people learn about their personalities, capabilities, intensity, determination, fears, hate, and even love. Some of us may be able to discover this dialog through agility drills or plyometric exercises. Often times, we can express ourselves through movement without even saying a word. We have an ability to explore through movement and be aware of our environment. Kretchmar (1993) explains that our dialogues are happening with mountains, valleys, rivers, and lakes.

PERSONAL PHILOSOPHY

Writing a personal movement philosophy is about developing our own movement ideas and practices, not just right resumes and curriculum vitaes. Kretchmar (1993, p. 264) brings up several useful questions to help us develop our own personal movement philosophies.

1. Does it bother you that some people think that science has answers for all of our problems and needs?
2. What if professionals relied wholly on taking the empirical turn and never took the reflective, philosophic turn?
3. Does it concern you that our schools emphasize reflective, sedentary education and generally neglect the active, embodied freedom to discover, explore, express, invent, and create?
4. Does it matter that many people badly underestimate the power of games, sport, and play to provide delightful and meaningful moments in life?

I want to find one of those places that movement is all about—places where words, word processors, and written sentences do not hold sway …

WHAT IS A PHILOSOPHICAL ATHLETE?

The concept of a philosophical athlete goes back to ancient Greece and a young wrestler name Plato who would be counted among the greatest thinkers of all times. Plato liked to describe philosophical dialogue in terms of wrestling moves and strategies. To him, and many others in ancient Greece, the philosophical struggle for truth was absolutely akin to the athletic struggle for victory.

Ancient Greek society provides a real-world model for how the synthesis of sport and philosophy can fuel the pursuit of personal excellence (*areté*) and the dynamic, thriving happiness the Greeks called *eudaimonia*. Education was for them, as it is for us, aimed at achieving a good and happy life. The problem is that, in modern society, we've retained our athletic programs but lost sight of the connection between physical education, excellence, and happiness.

As a college runner churning out lonely miles in the hilly country around Pocatello, Idaho, I thought little about such matters—beyond their obvious connection to the Olympic Games. For me, running was more than an escape from the books and lecture halls of the university, it provided a formidable challenge—a set of tangible standards by which I could test my personal mettle. I dreamt, like so many others, of an Olympic medal. But, as I ran along, imagining myself atop the Olympic podium, head bowed to receive a gold medal, my visions were less about the medal than about the "I" capable of winning it. The real task was to create the Olympian self, to cultivate the virtues—the discipline, the courage, the self-knowledge—I believed the Olympians had.

Now, as a college professor who never did stand on that Olympic podium, I can nevertheless say that sport brought me a long way toward being the kind of self I hoped would win a medal. I can say I was a philosophical athlete before I understood Plato or the Greek conceptions of excellence, education, and happiness. I kept my personal thoughts about sport and the intoxicating struggle for excellence to myself. Professors could not understand my devotion to sport, and coaches derided my emphasis on academics. I felt like I was the only person in the world who saw the connection.

It was not until I began working on my doctoral dissertation in the subject that I realized I was far from alone in my philosophic approach to sport. Reflected in my students' eyes, I see the desire for personal excellence shine through the frustration of being asked to articulate their reasons for participating.

Why sports? Initial responses to this question cluster around extrinsic rewards, such as wealth or admiration from others. Not every student who takes this class will become a philosophical athlete, but nearly all can gain a healthy perspective on the practice to which they devote so much of their time and energy.

A philosophical athlete focuses on the intrinsic rewards of sport, such as self-knowledge, ethical virtue, and learning to work with others as a team. The philosophical athlete knows that the greatest opponent is the self, the greatest challenge personal excellence, and the greatest reward true happiness.

The four characteristics of a philosophical athlete are as follows:

1. Values the sports experience as an opportunity to learn about him/herself as a person.
2. Takes responsibility for his/her actions, his/her attitudes, and the pursuit of meaningful goals.
3. Shows respect for him/herself, those around him/her, and the ideals of his/her sport.
4. Understands the values of his/her sports community and seeks to preserve them (Reid, 2002, p. 11).

CONCLUSION

The journey to becoming a true philosophical athlete cannot begin and end in the classroom. Only after successfully applying knowledge of these concepts can one truly be on their way to attaining philosophical athlete status. One's experiences with sports, movement, and classroom exposure could lead students down the path towards reaching this ultimate goal. For some, a key piece of the puzzle is actually defining what a philosophical athlete is and what a philosophical athlete does.

This text provides the opportunity to discuss and study the philosophical approaches to movement and physical activity and relate these to one's own experiences and ways of thinking. Philosophical athletes must possess a variety of qualities, some of which students may see in themselves. These qualities include (but are not limited to) having a deep understanding of the value of sports, viewing sports as a medium for self-expression, and having the drive to pursue personal success through sport.

Having a deeper understanding of sport and movement is crucially important to becoming a philosophical athlete. It is easy to think about sport superficially and just take it for face value. The philosophical athlete is able to connect with sport on a deeper level and get much more out of it. Many see basketball as more than just a sport that involves a basketball, a hoop, competitors, and a scoreboard. Basketball becomes a way of life; players live and breathe the sport, and it becomes part of their identity. During their time spent on the court, they often feel like time stopped. All their worries seemed to disappear by the all-too-familiar sounds of the game—squeaking shoes on spotless hardwood, the resonating drone of basketballs hitting the ground, and the blaring whistles of the referees. Even when they are not on the court, basketball follows them everywhere. They carry balls around school and dribble them as they go to their destinations. Many athletes do not realize that they are already on their way to becoming a philosophical athlete.

Another key quality of a philosophical athlete is viewing sports as a form of self-expression. When you watch someone perform a sport or movement activity, you are able to figure out a lot about that person just from observing. Understanding this concept allows you to use sport to better understand yourself and others. By attaining this knowledge, sports can be used to improve yourself as a person and identify possible weaknesses.

A philosophical athlete aims to pursue personal success through sport. Personal success does not necessarily mean winning, it involves success on a deeper level. For example, athletes that use performance enhancing drugs are typically in search of superficial success, whereas a philosophical athlete experiences success and happiness at a more meaningful level. Striving for a level of success that transcends the win–loss record or statistics from a game are what sets the philosophical athlete apart. This mindset allows for personal growth that goes beyond the sport that one is taking part in. Knowledge of this nature allows an individual to leave the sport he/she participates in and apply learning from that sport to other areas of their lives. It is hard to say that anyone can attain four characteristics of the philosophical athlete perfectly, but the value of striving toward this goal is immeasurable.

Many would suggest that movement shapes who we are as individuals. What we choose to do and how we choose to spend our time directly relates to how successful and happy we are. Humans are creatures that pursue knowledge through education and pursue happiness through effort. Sports give people the opportunity to test their capabilities by providing different rules and objectives to follow. These sets of rules also help to promote fair and proper human interaction within sports. These values are then learned and passed on to future generations. This knowledge and value system is a huge part of sport and is one of the best ways to educate someone on proper manners. Good sportsmanship is very important to our society and should be looked at as a way to teach youth how to interact properly. Sport also provides each team or player with an opportunity to overcome a challenge or an opponent to claim victory. This situation causes individuals to look at each event as an opportunity, to be better than their previous selves or better than their opponents. This creates a competitive atmosphere that often drives athletes to perform beyond their expectations. The gravity of competition and the satisfaction of victory drive the athlete to achieve excellence. In these ways, the pursuit of happiness through education and effort can be seen in sports. This ideal of perfection or greatness is broken down into four main issues that make up the philosophical athlete. The philosophical athlete participates out of pure joy for the sport and what the sport brings to him or her, also known as the love of the game. The athlete works hard to improve his or her performance on and off the field to help set a good example for his or her teammates, coaches, team, school, organization, town, or state, for everything he/she represents or that represents him/her. The athlete does not do it for the extrinsic factors like money or fame. The athlete does it, because he/she believes it is the right thing to do and it will make him/her a better person, because he/she loves to and knows he/she can make a difference.

The first characteristic of a philosophical athlete requires him/her to value the sport as an opportunity to learn about him/herself and improve. The athlete must want to be better every day and must work harder than everyone else to do so. The athlete must take each experience and analyze it from every angle. Critical thinking is essential when focusing on improvement. One must be able to weigh every option effectively and decide which action will yield the most positive outcome for the individual and the affected environment. Can we be one with the movement? Can movement describe us? This desire to be successful and happy in sports carries over to success in life. The battle may not be the same, but the end goal is.

A philosophical athlete must also accept full responsibility for his/her actions and attitude. He/she must show up with a positive, winning attitude every day. The athlete has to love the game and love his/her environment to truly give it his/her all and play with the right mindset. The athlete will usually take on a position of leadership, not only because he/she believes in his/her personal abilities, but also because he/she wants that leadership. The athlete believes he/she can help others perform better and influence them in a positive way, possibly off the field, as well as on. He/she must remember this too: always be humble and hungry. He/she has to want to succeed personally and as part of a team more intensely and frequently than anyone else. He/she must be hungry at all times. Those who do not personally identify with movement will claim this hunger is obsession. Do not be fooled, my friend. You just want it more than they do. Most people tend to sink right into the average or the "norm." Anyone that falls outside of that middle ground does not often go unnoticed. Change is uncomfortable and requires effort that most people are not willing to exert. This leads them to a conclusion regarding why someone else is living differently. More simply put, imagine you are doing everything within your power to be a great free-throw shooter. You stay in the gym for an extra 2 hours each day to practice, and often receive ridicule from your peers for being anti-social. Most average people do not want to put in the time or work to stand out and be great at something, so they instead find ways to bring you down. Jokes are made, and justifications are used to classify hard work as silly, especially when it is related to movement. If someone spends all day mastering algebra and then the entire next day mastering sprint mechanics, which one will receive more attention? Which one is deemed more important or credible? How are they even that different? Movement and biomechanics are almost entirely composed of physics, which is just another math-involved science. I digress.

Hunger must be accompanied by humility. Selflessness is the ultimate act of humility. The athlete must truly benefit from the intrinsic values of the sport, because he/she understands and respects the sport for what it is. He/she must find pure joy in the activity itself and the experiences shared with those around him. The athlete does not put himself/herself above anyone on the team. Though he/she aims to lead, he/she still walks alongside his/her teammates to provide the ultimate example of loyalty. The athlete takes on the toughest battles happily, because he/she enjoys the challenge, but also because he/she wants to protect his/her team and put it first. He/she is willing to sacrifice himself/herself for the good of others. The reward goes far beyond the satisfaction of the sport or the performance. The reward is knowing that you did everything you could for your fellow man and shared memories that will never be forgotten. You shared a bond, a unique connection with other people, and provided them with an example of how to act selflessly. Even if the season or the game is going poorly, you still maintain an honorable attitude. If you win, you are the first to shake hands. If you lose, you are the first to congratulate the other team. If the team falls down, you are the first person to get back up. You are the general; you are the hope that brings your team life. You have to get up, because you are not out there for yourself. Your humility prevents you from being blinded by arrogance.

Respect! A philosophical athlete must first respect himself/her. He/she understands his/her body is the vital performing instrument for sport so he/she takes care of his/her body. The athlete also understands that his/her mind is deeply connected with his/her body and his/her movement, so he/she must also take care of his/her mind. This requires a lot of discipline and knowledge, but it is a small price to pay for optimal performance within the athlete. The athlete must also respect the people he/she interacts with on and off the field. A true athlete composes himself/herself professionally in any public setting to show respect for his/her team and his/her community. He/she must also respect the sport itself and the values it teaches to others. Sports help people understand human interaction. They help develop manners and provide examples of polite behavior. They help different cultures share common interests and values by bringing people together. They also teach great values, like teamwork and humility.

Respect also ties into the values that sports help teach individuals and communities. Go down to Martin Stadium on a Saturday and tell me sports don't bring people/communities together. The excitement, the involvement, the entire event has an energy about it that you can feel. These feelings and emotions create last memories that impact lives. People literally behave differently just because of sports. In the simplest example, look at most Seahawks fans. They hate the 49ers just because they are rivals. This shared emotion brings Seahawks fans together when they might not have anything else in common. These experiences resonate through history, all the way back to the original Olympic games. Sports have brought people together and taught people how to interact for hundreds, even thousands of years. Movement has brought living things together and taught them how to live since the beginning of life. The upmost respect must be given to sports and movement because of how important they are in bringing people together under one common set of positive values. This leads back to people helping people. If there is any way to ensure sports continue to thrive and teach our youth how to live good and healthy lives, I would be glad to support it.

Some of the troubles faced in sport and movement are often very similar to issues in other aspects of our lives. They both involve the desire to win, improve, or overcome adversity. This desire must come from within individuals and is usually the factor that sets them apart. They desire something more from themselves. They work endlessly to perfect technique. They put in the extra time setting themselves apart. They take on extra responsibilities and put other people first. They respect the people they come in contact with and respect the sport in all its glory. They constantly strive for excellence and struggle for victory. Some athletes play sports to have fun, some play for the competition, some play for the interaction, but rarely do people play because they believe the sport makes them a better person. This extra effort builds confidence and satisfaction, which in turn, helps fuel inner desire even more. Movement can be used as way to test and challenge oneself, because facing adversity is followed by growth. The pursuit of excellence, the opportunity to be better than you were yesterday, is something that should be highly valued.

APPENDIX

Learning and Thinking Class Activities

LEARNING AND THINKING ACTIVITY ONE: THINK INNOVATIVELY

> Mind and body do not seem to act on another externally, as indeed they would if they were radically distinct entities. It is not accurate to say that a freestanding, independent mind tells the body what to do, or that a freestanding, independent body responds that it will or will not obey. Rather, when individuals think of purposes (like kicking goals in soccer) and supposedly tell their bodies what to do, they already are their bodies.
>
> (Kretchmar, 1994, p. 38)

Try to think in a different way. My mind and body are not two, but one, they work in harmony. Mind and body support one another for physical and mental purposes. Our physical activities and sport performances are more powerful when mind and body are working synergistically. Try to listen and pay attention to your body while exercising and/or working out. Are you ignoring your body by watching TV or listening to your iPod? Are you in tune with your movements? Reflect on how you view your mind and body. Do you view them as separate entities or as working together in harmony?

LEARNING AND THINKING ACTIVITY TWO: NEW TERMS

Avoid referring to bodies as something that people have, own, or otherwise bring along with them. When you use such language, this has the effect of distancing you from your physical nature. Things that you have or own can

> be left behind. But obviously you cannot do that with your physical nature. Thus, you do not just have a body; you are your body.
>
> (Kretchmar, 1994, p. 40)

"I am my body!" All movement and physical involvements are experienced through the body. How do you view yourself? If you can use three words to describe yourself, what would they be? What are your strengths? What are your weaknesses? What motivates you to be physically active? Are you aware of your extrinsic motivation? Is it hard to develop your intrinsic motivation to movement?

LEARNING AND THINKING ACTIVITY THREE: BODY AS AN OBJECT

> Don't pretend that you can manipulate bodies—work on them, train them, educate them—without affecting the whole person … Heart rates do not get produced in a vacuum. They come with pain, and accomplishment, and interest, and boredom, and love, and hope, and virtually any other affect that can be thought of. They come attached to human purposes and embedded in human stories. They are related to what people have been, what they are now, and what they hope to become.
>
> (Kretchmar, 1994, p. 41)

How do you refer to your body? Is your body an object that can be manipulated? Do you practice a mind-over-matter philosophy? Saying, "just do it" or "make your body your machine," suggests to our body being objectified and trained. Try to avoid referring to your body as "it." How can you change the way you refer to your body? What are some ways you hear people objectify their bodies?

LEARNING AND THINKING ACTIVITY FOUR: WORK OUT OR PLAY?

> Sensuous joy of movement, satisfaction over accomplishment. Speak of movement activities, fun, enjoyment, play … not in spite of the fact they are "physical and sensuous," but *because* they are so. Workouts should be more often become PLAY OUTS.
>
> (Kretchmar, 1994, p. 100)

When you exercise, is it fun? People do not say, "I could have exercised all day long," but people say, "I could have danced or golfed all day long." How can you make your workouts and exercise feel more like play? How can you make them fun? Do you share your physical activity experiences with others?

LEARNING AND THINKING ACTIVITY FIVE: PLAY IS SELF-DISCOVERY!

> If you are doing poorly in any "play" (sport) stop thinking of success as a product of the mental and error the product of the physical. No more—the body failed you. As we improve our skills, we experience a lived freedom to search, explore, invent, express, and create. ... This freedom is exhilarating just as the freedom is experienced when a proposition or fact erases ignorance. We like ourselves as free, competent, perhaps even powerful. We want to return to the court or pool or dance floor or exercise hall—wherever it is we experience ourselves like this.
>
> (Kretchmar, 1994, p. 168)

How can you make your exercising sessions feel more like play? What are some things you can discover about yourself through play and movement? Do you feel successful when you participate in physical activities?

LEARNING AND THINKING ACTIVITY SIX: WHAT IS THE GOOD LIFE?

> The Good Life refers to an overall life condition and set of experiences that we regard as desirable. While most everybody aims at good living, there is considerable disagreement about what exactly it is. There are probably hundreds of ways to achieve something called the good life, and there are undoubtedly many patterns that are comparably good.
>
> (Kretchmar, 1994, p. 111)

> A good life should entail pleasure and fun, which is almost universal. Movement should bring meaning to your life. The good life should be meaningful, and physical activity should bring meaning.
>
> (Kretchmar, 1994, p. 224)

The good life should incorporate pleasure and fun. Physical activity, sports, and exercise should bring meaning to our lives. Movement experiences are significant and should be appreciated for their own sake. We should find purpose to movement and be able to develop our own stories. Survival and long life may not be sufficient to assure the good life. What is the purpose of exercise? How would you define "the good life?"

LEARNING AND THINKING ACTIVITY SEVEN: DO WE PARTICIPATE IN MOVEMENT ACTIVITIES FOR SKILL OR KNOWLEDGE?

> Propositions about significant living are not difficult to locate. The Bible, the Koran, scriptures, hundreds of sages of wise people [sic], and thousands of pop psychology and philosophy books can supply them … The fact is that we grow into stories and meaning more than we encounter them as foreign propositions or theories. Cultural traditions, hobbies, dances, games, habits, crafts and other activities point us in some directions and away from others. The skills we learn tell us implicitly that it is important to do this and not that. Moreover, these activities come loaded with values—with etiquette, with ways of behaving, with right attitudes and so on. By learning play and game skills, we grow into rights and wrongs, values and disvalues, things that are important and other things that are not valuable … Games and play resonate with the dominant messages of our time and place, and this builds a compass into our being that tells us the directions in which we should develop our personal stories.
>
> (Kretchmar, 1994, p. 167)

Do you do physical activity for fitness, pleasure, skill, or knowledge? Which one is the most important—can you rank them? Should skill and pleasure be ranked before anything else to promote a meaningful life? Why do we move? How can our physical activities contribute to our skill and knowledge growth? What are some activities that can contribute to skill and knowledge development?

LEARNING AND THINKING ACTIVITY EIGHT: SHARING OUR PHYSICAL ACTIVITY EXPERIENCES

> Pleasure is influenced by familiarity. Activities that were once a great deal of fun sometimes grow stale. People have mood swings that influence the degree of enjoyment that is experienced. External factors, like the attitudes of other participants in a group activity, can dramatically affect the level of enjoyment … Thus, if continued participation or repeated involvement were to be based only on pleasure, it might be on very unsure footing.
>
> (Kretchmar, 1994, p. 146)

Pleasure should be the most important value to physical activities. It is what gets people hooked to the games and play. Pleasure motivates people to be active. How can you maintain the enjoyment of your physical activities?

LEARNING AND THINKING ACTIVITY NINE: INTELLIGIBLE EXPERIENCES

> Perhaps the most fundamental and customary experience of a coherent life comes with developing and living a story. As we mature, move through school, prepare for our life's work and play, make commitments to mates or choose to remain single, and start families and careers, we continue to define our stories and achieve, hopefully increasing degrees of meaning, satisfaction, and at-homeness with our choices. We refine our goals and gain a good sense of what fits and what does not.
>
> (Kretchmar, 1994, p. 131)

Significant movement experiences build a meaningful and pleasurable life stories. Can we define our lives through movement? Exercise should be the habit to break. Movement should be something we do forever. Movement should be a joy to look forward to. "I get to exercise and workout today," not "I have to work out today." "My day begins with play!" "I look forward to movement!" "I look forward to being creative!" "I incorporate movement into everything I do!" How can you develop our own movement stories?

LEARNING AND THINKING ACTIVITY TEN: IDENTIFYING OURSELVES THROUGH EXERCISE, PHYSICAL ACTIVITY, SPORTS, GAMES, AND PLAY

> Discovery. Human movement is a dialogue between persons and a spatiotemporal world. The dialogue is given life by purposes—to play, to win, to score, to kick, to show. As the dialogue unfolds, discoveries typically trip along one after another. People learn about themselves—their personalities, their capabilities, their intensity, their determination, their generosity, their fears, their tenderness, their prickliness, their capacity for love, their potential for hate. This information does not come inscribed on parchment. It comes as human beings jump, through their victories and defeats, when they swing or pirouette, as they fall or dive, while they pass to a teammate or get shut out of an offensive scheme. This process of discovering can be valued for its own sake.
>
> (Kretchmar, 1994, p. 195)

What if you try to describe yourself as a runner, where you no longer see yourself as a slave to running, but running is your identity? Running is who you are. You define yourself through running. It becomes joy, not work. Every single time you go for a run, it is significant and special. Everyone should want to achieve the values of movement and share them with their friends and family. What would you say to someone who asks you if you exercise and/or workout? Would you rather identify yourself as a mover in activity or a worker in exercise?

LEARNING AND THINKING ACTIVITY ELEVEN: INTRINSIC VALUE OF PHYSICAL ACTIVITY

> Pleasure often occurs in environments where something of a spell has been cast over its participants they are give to the experience—so given, in fact, that they are not sure why they spent so much time there. When asked after the fact why they gave so much energy to dancing, playing field hockey, or riding a bicycle, for instance, these players will often say simply that they experienced a great deal of pleasure. Or more likely, they will just say it was fun.
>
> (Kretchmar, 1994, p. 168)

> It may be true that those with the greater skill levels may have more frequent or deeper experiences of play, but skill is not needed for play to take over people's lives. Children with skant knowledge and very few skills (motor or otherwise), have rich and memorable play experiences.
>
> (Kretchmar, 1994, p.171)

Have you ever experienced the true joy of movement? The true beauty of movement? Beautiful means pleasant to the senses (sight, sound, smell, and touch). For example, when I move, my eyes provide the beauty of the river, the tress, the sky ... When I move, I focus on the rhythmic pattern of my feet, I focus on my breathing and the beat of my own heart. Even when I move in the city, the noise brings me joy. I taste the salt of my sweat. I feel my eyes burning, but the movement itself still gives me joy. My muscles fatiguing still gives me joy. This movement is effortless. Try to immerse yourself in a culture of your favorite movement. Watch your favorite sport, watch your favorite game, watch your favorite play. How do you allow yourself to experience the joy of your physical activity? Be intriguing, be fascinated, and be captivated about your physical activities!

LEARNING AND THINKING ACTIVITY TWELVE: EXERCISE/WORKOUT DILEMMAS

> While everything we do carries some sort of meaning, everything we do does not have to be meaningful. Sometimes in the midst of my frenetic life, I find myself needing a vacation from meaning.
>
> (Kretchmar, 2001, p. 25)

Often times, exercising is initiated for health reasons and sometimes because it is fun. Usually, that fun wears off. We may begin to exercise, because we want to feel and look good, or be fit. But, that need eventually disappears. What if we became so used to our physical activities that we did not need a reason to continue. We participate in physical activity, because it is who we are and what we do. What do we need to do to establish meanings and that intimate relationship with our physical activities?

LEARNING AND THINKING ACTIVITY THIRTEEN: DEVELOPING MOVEMENT STORIES

> A life story has a beginning, middle, and an end, and all three are related to one another. A story leads someplace. Its characters have roles to fulfill, work to do, and celebrations to hold, and love to experience.
>
> (Kretchmar, 1994, p. 131)

Why write a movement story? Why write about our mind/body connection? We can share these stories with individuals seeking to develop their own. It will be up to you to develop and maintain your own movement story that will continue through your life. How do you see adapting yourself to life after college?

REFERENCES

Anderson, D. R. (2002). The humanity of movement or "it's not just a gym class." *Quest, 54*, 87–96.

Blumhagen, D. W. (1979). The doctor's white coat. *Annals of Internal Medicine*, 91, 111–116.

Corbin, C. (2002). Physical activity for everyone: What every physical educator should know about promoting lifelong physical activity. *Journal of Teaching in Physical Education*, 21, 128–144.

Dicker, G. (1998). *Hume's epistemology and metaphysics*. New York, NY: Routledge

Eitzen, D. S. (1984). School sport and educational goals. In D. Eitzen (Ed.), *Sport in Contemporary Society: An Anthology* (199–202). New York, NY: St. Martin's Press, Inc.

Forencich, F. (2006). *Exuberant animal: The power of health, play and joyful movement*. Bloomington, IN: AuthorHouse.

Husserl, E. (1902, 1913, 1962). *Ideas: General introduction to pure phenomenology*. (W. R. Boyce Gibson, Trans.). New York: Collier Books.

Husserl, E. (1923, 1961). *The Paris lectures*. (P. Koestenbaum, Trans.). The Hague: Martimus Nihoff.

Husserl, E. (1975). *The Paris lectures* (2nd ed.). (P. Koestenbaum, Trans.). The Hauge, Netherlands: Martinus Nijfoff.

Husserl, E. (1989). *Ideas pertaining to a pure phenomenology and to a phenomenological philosophy*. Boston, Massachusetts: Kluwer Academic Publishing.

Huzinga, J. (1950). *Homo ludens: A study of the play element in culture*. Boston, Massachusetts: Beacon Press.

Jackson, F. (1991). What Mary didn't know. In D.M. Rosenthal (Ed.), *The Nature of Mind* (392–394). New York, NY: Oxford University Press

Kant, I. (1993). *Grounding for the metaphysics of morals: On a supposed right to lie because of philanthropic concerns.* Indianapolis, IN: Hackett Publishing Company, Inc.

Kretchmar, R. S. (1994). *Practical philosophy of sport.* Champaign, IL: Human Kinetics.

Kretchmar, R. S. (2005). *Practical philosophy of sport and physical activity.* Champaign, IL: Human Kinetics.

Landgrebe, L. (1977). *Phenomenology as transcendental theory of history.* Notre Dame, IN: Notre Dame University Press

Merleau-Ponty, M. (1962). *Phenomenology of perception.* New York, NY: Humanities Press.

Merleau-Ponty, M. (1964). *The primacy of perception.* J.M. Edie (Ed.), Evanston, IL: Northwestern University Press.

Metheny, E. (1968). *Movement and meaning.* New York, NY: McGraw-Hill Book Co.

Metheny, E. (1972). The symbolic power of sport. In *Sport and the Body: A Philosophical Symposium* (221–226). Philadelphia, PA: Lea and Febriger

Metheny, E. (1979). The symbolic power of sport. In E.W. Gerber and W.J. Morgan (Eds.), *Sport and the Body* (2nd ed.) (p. 231–236). Philadelphia, PA: Lea and Febiger.

Mijacevic, D. (2013). *Where soul meets body: It's not just a workout: Lived sport and exercise experiences.* Saarbrücken, Germany: Lap Lambert Academic Publishing.

Plato. (1985). The separation of body and soul. In W.J. Morgan and K.V. Meier (Eds.), *Philosophic Inquiry in Sport* (68–72). Champaign, IL: Human Kinetics.

Polanyi, M (1969). The two cultures. In M. Polanyi and M. Green, *Knowing and being.* Chicago, IL: The University of Chicago Press.

Reid, H. L. (2002). *The philosophical athlete.* Durham, NC: Carolina Academic Press.

Sheets-Johnstone, M. (1999). *The primacy of movement.* Philadelphia, PA: John Benjamins North America.

Slowikowski, S. S., and Newell, K. (1990). The philology of kinesiology. *Quest,* 42, 279–296.

Stoll, S. K. (1980). *The use of Merleau-Ponty's phenomenological method to describe the human movement forms of the third dynasty of UR.* (Unpublished doctoral dissertation). Kent State University, Kent, Ohio.

Stoll, S. K. (2009). *Sports ethics.* Retrieved from http://www.educ.uidaho.edu/stoll/PEP570/finalform.htm

Suits, B. (1972). What is a game? In *Sport and the Body: A philosophical Symposium* (16–22). Philadelphia, PA: Lea and Febiger.

Todd, W. (1979). Some aesthetic aspects of sport. In R.H. Fox (Eds.), *Philosophy in Context* (8–21). Cleveland, OH: Cleveland State University.

Twietmeyer, G. (2012). What is kinesiology? Historical and philosophical insights. *Quest,* 64, 4–23.

ACKNOWLEDGMENTS

This book is the product of efforts of many people. I have attempted to credit those individuals whose ideas and arguments I have borrowed. To any I have failed to mention, I offer my sincere apologies. Particular gratitude should be directed to Dr. Sharon Stoll, a friend, a mentor, and the individual who taught me a great deal about movement philosophy and philosophy in general. More important, Dr. Sharon Stoll provided an example of academic encouragement and fairness. In short, a profile of a fair athlete/person. Also, thanks to Kristina Stolte, my senior field acquisition editor from Cognella, Inc., and Sarah Wheeler, my project editor; to my colleagues at Washington State University, who have taught, encouraged, and stimulated me over the years; to my KINES 314 (Philosophy of Human Movement) students at Washington State University, who have provided some interesting ideas, thoughts, and dilemmas described in this book; and finally, to all of my friends and family who have supported me through this project.